Coastal and estuarine management

There are many ways in which humans can bring about change to coastlines and estuaries. Many of these changes are caused by engineers or coastal managers, but others are caused by casual visitors, holiday-makers, and people who just visit the coast for a day out by the sea. Very few of these changes are intentional, but tend to arise as a result of the many activities which occur within the coastal zone.

This book presents the various ways in which such changes and impacts can occur. Having described the basic operational processes, the book goes on to look in detail at the main spheres of activity which impact on the coastline. Coastal defences are perhaps the greatest causes of change and problems. Building a sea wall can cause problems elsewhere along the coast by cutting off sediment supply, thus causing beaches to erode. Other defences aim to stop sediment moving from a beach, such as in tourist areas where the presence of a beach is vital. However, we must start to question the impacts which beach defences cause elsewhere along the coast, and ask whether there are better ways of achieving the same end result.

Having considered the various problems, each section considers the alternatives. The case is argued that, in many areas, there are better alternatives to traditional sea defence techniques. Soft engineering allows the coast to be more natural, and function in a more natural way. In effect, by using soft engineering techniques, the coast can be protected more cheaply than it is at present, and with considerably less long-term upkeep. This begs the question as to why these techniques are not more widely used, a problem discussed in the final chapter, which considers the whole coastal management issue.

Peter W. French is a lecturer in Physical Geography at Lancaster University, with a particular specialism in coastal processes and management. He has published work relating to many aspects of human coastal impact, including salt marsh pollution, erosional problems of borrow pit construction, human causes of coastal habitat loss, and causes of salt marsh loss in the UK.

Routledge Environmental Management Series

This important series presents a comprehensive introduction to the principles and practices of environmental management across a wide range of fields. Introducing the theories and practices fundamental to modern environmental management, the series features a number of focused volumes to examine applications in specific environments and topics, all offering a wealth of real-life examples and practical guidance.

MANAGING ENVIRONMENTAL POLLUTION
Andrew Farmer

COASTAL AND ESTUARINE MANAGEMENT
Peter W. French

Forthcoming titles:

ENVIRONMENTAL IMPACT ASSESSMENT
A. Nixon

WETLAND MANAGEMENT
L. Heathwaite

COUNTRYSIDE MANAGEMENT
R. Clarke

Coastal and estuarine management

Peter W. French

London and New York

First published 1997
by Routledge
11 New Fetter Lane, London EC4P 4EE

Simultaneously published in the USA and Canada
by Routledge
29 West 35th Street, New York, NY 10001

Typeset in Ehrhardt by Keystroke, Jacaranda Lodge, Wolverhampton
Printed and bound in Great Britain by Redwood Books, Trowbridge, Wiltshire

British Library Cataloguing in Publication Data
A catalogue record for this book is available from the British Library

Library of Congress Cataloguing in Publication Data
French, Peter.
        Coastal and estuarine management / Peter French.
        (Routledge environmental management series)
        Includes bibliographical references and index.
        1. Coasts. 2. Estuaries. 3. Coastal zone management.
        4. Estuarine area conservation. I. Title. II. Series.
        GB451.2.F75   1997
        333.91'715 – dc21 97–14868

ISBN 0-415-13758-6 (hbk)
ISBN 0-415-13759-4 (pbk)

Contents

Plates

Figures

Tables

Boxes

Preface

This book is intended to provide undergraduates studying coastal and estuarine processes with a solid grounding in the ways in which human activities impact on these environments. These impacts take many forms; some are direct, and some are indirect. Throughout history, humans have interfered with coasts and estuaries, with, until fairly recently, little understanding of the wider implications of what they are doing. This book aims to inform readers of these wider issues, and, in addition, of the alternative approaches which, while not necessarily being impact free themselves, do provide a better and more sustainable approach to managing the coastline.

The key words throughout are *sediments* and *pollution* because every impact can be brought down to one of these common denominators. Pollution can affect wildlife, habitats, and our enjoyment of the coast. Sediment is the basic building block that determines whether a coast is accreting, eroding, or stable.

Taking these two common factors, the book looks at the main issues affecting the coastline. Having seen how the coast and estuaries operate, and the type of habitats which result from the various forms of coastal process, we look in detail at coastal defences, industrial development, and tourism. It is important to consider, however, that it is possible to affect coastal processes indirectly, by interfering with catchments, land use, and offshore conditions. In recognition of this, we also discuss the impact of distant activities.

In a world where coastal management methods are rapidly developing, and much of the coast is becoming affected by a variety of coastal plans, the issue of coastal management is the subject of an important concluding chapter to look at how all the impacts discussed are being managed.

Before we proceed further, it is necessary to develop an understanding of some of the key concepts which are used in this book. The coast is not static, and is constantly changing in response to waves, winds, and tides. It is important to realise that what we see now is only a snapshot in time, and there is no reason on earth why humans should adopt a policy of insisting that the coast should remain as it is. Such an attitude is only a prelude to serious problems in the future. In addition, as far as we are concerned here, the coast can be considered as serving two main roles. First, it is a geomorphological unit which stops waves and keeps the sea off the land. Second, all features that we see on the coast, be they beaches, marshes, dunes, spits, etc., are sediment stores. These stores have been left temporarily, at a time when they are surplus to requirements. When conditions change, the sea will utilise these stores. If humans have developed these areas, then problems occur and conflict between human society and the sea results. It is these problems which we look at in detail in the following pages.

Acknowledgements

The preparation of this book has involved the support and encouragement of a lot of people. The established authors in the Geography Department at Lancaster provided some very important tips, which probably saved a lot of time in the long run. Many of my fellow physical geographers also provided additional support, not all of which was alcohol related. The encouragement of friends and family has been invaluable, especially at times of computer hardware failure.

I have been grateful for the support of Sarah Lloyd, my editor at Routledge, for being supportive and quick with advice when needed. The figures also owe a great deal to the departmental cartographer, Claire Jeffrey, who has converted my scribblings into something useful. In addition, the useful and constructive comments of an anonymous referee are gratefully acknowledged.

My thanks are further extended to the following publishers for their permission to use copyright material:

Figures 1.6 and 2.2 from Hansom, J.D. (1988) *Coasts*. Reprinted with permission of Cambridge University Press, and the author.

Figure 3.1 from Davidson, N.C. *et al.* (1991) *Nature Conservation and Estuaries in Great Britain*. Reprinted with permission of the JNCC and English Nature, as well as the author.

Figure 4.6 from the WWF-UK *Marine Update* no. 25. Reprinted with permission of WWF-UK, and the author of this report, Dr Peter Dyrynda.

Chapter 1

Introduction to estuarine and coastal systems

Introduction

There are very few environments left on planet Earth where human impacts are not manifest in some form. Many human activities bring about changes which result in some degree of artificialness to many environments. While it is true that some of these activities will bring change which is to an area's benefit, such as the creation of new habitat or increased environmental stability, in many cases the environment suffers. However, whether change is beneficial or not, an environment, once altered, ceases to be natural, and the natural processes which operate in that environment are affected by unnatural conditions.

Coasts and estuaries are typical environments in which human impacts have led to a whole range of changes with considerable variation in their degree of impact. The coast is typically a highly populated area (for example, Goldberg (1994) states that 50 per cent of the world's population lives within 1 km of the coastal zone, and this population is predicted to grow at a rate of 1.5 per cent over the next ten years). In addition, many people also see the coast as a place to spend their leisure time, taking advantage of the beaches and attractive scenery; indeed, some people may associate the coast purely with such activities. Also, coasts and estuaries are seen as places to develop industrial or port sites, taking advantage of abundant water supplies, transport links, and sheltered areas for loading and unloading ships. As a result, the coastal zone (defined in many ways by different researchers, but generally taken to refer to the area of land which is influenced by the sea, and vice versa, and thus including areas of land and sea affected by coastal activities) experiences intense pressure and demands from various sectors of the community. This pressure has the potential to cause change and environmental degradation, if not carefully managed.

Historically, human interference within the coastal zone has had the intention of protecting financial interests, increasing earning potential, or providing places to relax and spend leisure time. It has also been traditional to carry out these modifications with humans seen as the central focus of the 'development tree'; that is, all changes are seen in the light of benefits to humans (humanist approach) (Figure 1.1). However, if we substitute 'Humans' in Figure 1.1 by 'Environment' (Figure 1.2) this set of 'benefits' can be seen as impacts on the environment, and becomes less appealing from a non-humanist perspective. From an environment point of view, for example, what is good for tourism can be harmful for habitats. Coastal defences, while necessary for infrastructure protection, produce significant impacts on coastal sediment budgets, and can merely result in the transfer of an erosion problem elsewhere along the coast (Chapter 3). The logical result of this argument is that any activity which occurs along the coast has the potential to be environmentally damaging. From this we can clearly see that, ideally, the only way to ensure that human action does not cause damaging

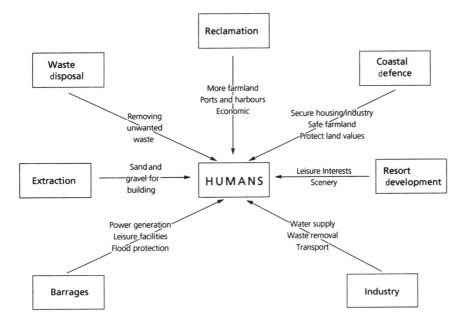

FIGURE 1.1 Development of the coast as seen from a human perspective

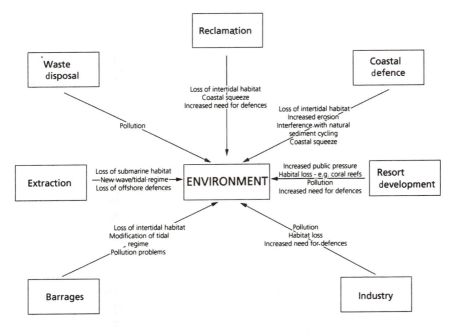

FIGURE 1.2 Development of the coast as seen from an environmental perspective

effects along the coast is not to undertake any activities at all. However, given that we cannot (and should not) stop human activities at the coast, it is necessary to operate under a system of management, by which we control the activities which do occur and manage them in such a way as to minimise the detrimental effects on the environment, and to ensure that important habitats are left alone. This book addresses many of the aspects which cause management problems along coastlines. Individual chapters will investigate coastal defences, land claim, industrial developments, tourism, and habitat exploitation and analyse these in the context of natural processes versus anthropogenic actions. A common thread linking all these aspects, however, is that it is the ways in which humans use the coast which cause the problems, not the natural processes. Clayton (1995) states that if humans did not seek to make use of the coastal zone, there would be few problems as a result of the occurrence of coastal erosion; that is, erosion, and the subsequent need for defences, is considered a problem only because human activities have occurred which make land too valuable to be allowed to erode.

The more of the coast which can be allowed to erode naturally the better, because in such a situation, natural processes can operate under natural conditions. The problems faced today largely stem from the fact that the coastal zone is too much in demand. As a result of this increasing demand, it is imperative that some form of management is put in place. Chapter 7 will discuss this in detail, and demonstrate that in respect of management, time is very much of the essence, as exploitation of the coastal zone is continuing and increasing. Coupled with this, it must also be remembered that processes of human-induced change are continuing against a background of long-term environmental change, most notable of which is sea level rise. Indeed, some researchers would argue that, given the role of global warming in sea level rise, even this process is related to human impact. However, the causes are more complex than this, with other factors, such as isostatic land movements, being as important, if not locally more so. In our subsequent discussion of human impact on coasts and estuaries, we will inevitably touch on this subject area, although we will avoid entering into the debate of causes of sea level rise, as these are well addressed in other dedicated texts (for example, see Bird, 1993; Bray et al., 1992; Tooley and Jelgersma, 1992).

A second process which has a direct bearing on management (Chapter 7), and also one which can further magnify the effects of some human actions, is the change in wind–wave climate. Again, this area of research is one of intense debate because although evidence suggests that some seas and oceans, such as the north-east Atlantic and the North Sea, are becoming rougher (Bacon and Carter, 1991; Carter and Draper, 1988; Hoozemans and Wiersma, 1992), it is argued that this may be purely a short-term fluctuation in measurement, as opposed to a long-term change. Equally though, increased water depth due to sea level rise and associated increases in wave base can also produce a similar effect of increasing wave activity inshore.

The combination of human-induced change and accelerated sea level rise is leading to the realisation that there is a growing body of evidence to indicate that coastal erosion is becoming a serious threat along many parts of the world's coastline. Some coastal researchers argue that there appears to be a pattern emerging in erosion

(Bird, 1985, 1987; Paskoff, 1987), as a result of a study by the International Geographical Union between 1972 and 1976. This study included all countries with a sea coast, and concluded that the only areas of coast which still show rapid accretion are either where large rivers supply abundant sediment to the nearshore zone, or where coasts are experiencing isostatic uplift (i.e. relative sea level fall). Leaving these two exceptions aside, many other coasts are showing signs of adjustment to compensate for 'natural' and anthropogenic changes. Many examples of these changes appear in the literature, and are cited in subsequent chapters.

The causes of this erosion are also various. In Italy, intensive cultivation and grazing prior to this century increased sediment yield to the coasts to allow accretion and stability (Bird, 1985; Tilmans *et al.*, 1985). Subsequent decreases in sediment yield, partly due to changes in land use and partly due to increased soil erosion control measures, mean that these large sediment accumulations can no longer be maintained, and so a state of net erosion has set in. Similarly, much of the Mediterranean coast has experienced erosion due to the retention of sediment behind dams. This problem, although serious along the coastline in general, is especially manifest in deltas. The Rhône, Po, and Nile deltas are all suffering erosion through sediment starvation. An interesting contrast occurs in the eastern Mediterranean where, because there has been less extensive human intervention, coastal erosion has been less marked (Hanson and Lindh, 1993).

Along the Atlantic coast, erosion rates of 150 m between 1952 and 1969 were recorded in the Brittany region of France, and 700–800 m between 1881 and 1922 along the coast of the Bay of Biscay. Further problems along the Dutch coast have been reported by Louisse and Kuik (1990) and by Rijkswaterstaat (1990). The North Sea coast of Scandinavia has also seen accelerated erosion with part of the Swedish coast losing 150 m between 1820 and 1869 and one locality, Löderup, losing 20 m in one year (Hanson and Lindh, 1993).

In the USA, all thirty coastal states are experiencing some degree of erosion, much of it due to the interference of humans in natural coastal processes as a result of development. Figure 1.3a demonstrates the severity of this erosion diagrammatically. Part of the reason for this development is an increase in population. In 1980, 50 per cent of the US population lived within 75 km of the coast. By 2010, this figure is expected to reach 75 per cent, representing a considerable increase in demands for facilities, and subsequent impingement on natural coastal resources (Hanson and Lindh, 1993).

This problem of population increase at the coast is not unique to the USA, however. In 1983, a study of the coastline of western and central Africa (Quélennec, 1987) showed that human influences were one of the prime factors behind erosion. In Ghana, erosion rates of 6 m yr^{-1} occurred between 1964 and 1975 and maximum rates of 25 m yr^{-1} have been recorded in Togo. More specifically, erosion induced by harbour construction has been measured as being between 4 and 30 m yr^{-1} to the east of Lagos, and around 2 m yr^{-1} near Monrovia, Liberia, due to the halting of longshore drift (Hanson *et al.*, 1984). All these studies appear to support a link between human activity at the coast and increased erosion. The precise causes of this erosion are covered in subsequent chapters.

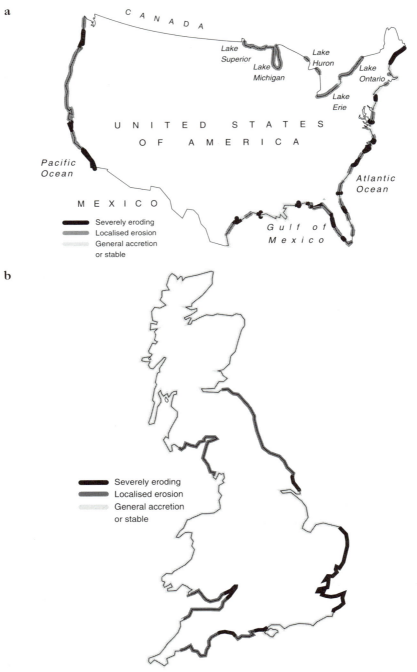

FIGURE 1.3 Generalised trends of coastal erosion and accretion in the USA (a) and the
UK (b)
Sources: (a) Modified from Hanson and Lindh, 1993; (b) based on data from Pye and French,
1993a

In the UK, a study similar to that for the USA (Figure 1.3a) has revealed distinct geographical zonation in the erosion and accretion trends around the English coast (Pye and French, 1993a, 1993b). Figure 1.3b reveals a general trend of severe erosion around the south-eastern coast, slight erosion in the north-east and south-west, and general accretion in the north-west. The problem in the south-east is largely due to human interference in the intertidal zone and sea defence construction as a result of increased population and increased demand for coastal leisure facilities. Such demands are not restricted to the UK, however. Any coast which has an appeal to tourists is under the same threat, and, while the above example relates to the UK, similar examples can be found relating to coasts of the USA (Carlson and Godfrey, 1989; McAtee and Drawe, 1980), Europe (Ballinger *et al.*, 1994; Charlier and de Meyer, 1991), and Australia (Kenchington, 1993; Yapp, 1986), as well as many other parts of the world. In south-eastern England these problems are made more serious by the isostatic movement of the land, which means that as well as development, 'natural' processes such as sea level rise are operating in conjunction with sinking land, thus increasing the relative movement of sea level. The UK problem is serious enough to cause English Nature, the main English conservation body, to initiate a major project to halt the loss of intertidal habitats. Pye and French (1993a) studied the key habitats around the English coast and determined the predicted loss of each over the next twenty years. Box 3.1 in Chapter 3 details these findings.

Using the basic arguments outlined above, this book will provide the reader with an account of the ways in which humans have influenced and brought about change to the coastal and estuarine environments similar to those detailed, what effects this has had, how we are now beginning to realise the problems which have been caused, and the methods being employed to solve them. It will then provide a structured management framework for the coast and estuarine environment.

The structure of this book

This chapter (Chapter 1) will introduce the reader to the types of coastline and estuary we find, and how these different types can be influenced in different ways by human activity. In Chapter 2, the physical regime of coasts and estuaries will be introduced. We shall look at formational processes, such as the effect of tides, waves, and wind, and how these processes affect the movement of sediment. By doing this, we see how the system operates, and thus, in succeeding chapters, we can see how, by interfering with these processes, we can produce a whole new set of problems which need solving, often causing greater impacts than would be the case had natural processes been allowed to proceed unchecked.

Following this, Chapter 3 will introduce some of the key activities, such as land claim, coastal defence, and port development, which have, historically, taken large quantities of intertidal land with resulting modification to the environment. This is followed in Chapters 4 and 5 by an examination of the development of industry and tourism, and then Chapter 6 will look at ways in which humans have caused changes at the coast by action remote from it, such as activities within the river catchment causing changes in sediment supply, and the winning of offshore sediments, which can result in increased wave activity.

These main aspects of human impact having been considered, the final chapter will look at how many of the problems raised have been and can be tackled, with respect to environmental protection and conservation.

Types of estuarine and coastal environments

If you visit the same piece of coast on different occasions, you will probably notice that it changes. These changes may be many and varied, and occur on a variety of scales (Table 1.1). For example, there may be a new coastal development such as a hotel or arcade, or you may feel that there is less sand on the beach than before, or that there has been a large cliff fall. It is a fact that these changes are going on all the time. Whenever the tide comes in and goes out, it brings with it a series of waves which work on the coast, shaping it and reworking it (Table 1.1). These processes can largely operate with little worry for humans. The time that problems start to occur is when this constant remodelling of the coast starts to affect human society – cliff erosion is threatening a house, a beach is eroding so the tourist potential of an area is reduced, a salt marsh is eroding and threatening the loss of important wildlife habitat or farmland. It is at this point that human intervention can start and the natural processes cease to function under natural conditions. The natural thing for such coasts to do in these circumstances is to erode. By interfering, we can stop this natural process and force the coast into an unnatural state. When in such a state, the coast will continue to attempt to reach an equilibrium state, and in so doing may erode sediment from elsewhere, or from the same locality, causing beach loss.

While changes to the coastline are continuing all the time, the coasts and estuaries of the world in which these processes operate are also extremely variable, and many texts exist solely concentrating on these in their own right. It is beyond the scope of this text to undertake an in-depth study of coastal landforms, although the next chapter will deal with coastal processes as they are affected by the activities of humans. For detailed information on coastal morphology, the reader is referred to other suitable texts, such as Carter (1988), Hansom (1988), and Pethick (1984).

In its most general terms, the coast represents the boundary between land and sea and can be seen as serving two functions. First, the coast is there to stop waves, and second, it serves as a store of sediment. It is important that the coast should be seen in this way. Coastal landforms are not permanent features, they are transient features which accrete and erode subject to the environmental conditions at the time. Coasts and estuaries are highly dynamic environments, and constantly change in response to natural forces on a variety of time scales (Table 1.1, Figure 1.4). For example, sediment grains can be moved on a beach or mudflat whenever waves lap against the shore; a storm can change the profile of a beach in a few hours; a winter's rough seas can modify a soft cliff profile over a couple of months; or sea level can rise and fall over thousands of years. Thus it is important to keep in mind that the coast which we see today is not static and frozen in time, and that there is no reason why

TABLE 1.1   The time scale of coastal changes in relation to both absolute and human time scales

Absolute time scale	Human time scale	Coastal processes
Millennia		Response of sea level to glaciation
Centuries	Shifts in settlement and industry	Historic coastal change – loss of towns and villages
Decades	Coastal engineering and protection	Formation and loss of habitats – marshes, dunes, etc.
Years	Coastal engineering, management plans	Effects of protection works, longshore drift
Months	Impacts of tourism	Seasonal adjustments, shore profiles
Weeks	Impacts of tourism, emergency coastal protection works, extraction	Shore profiles, spring–neap tidal cycles
Days	Emergency flood protection works	Storm surges, defence breaches
Hours	Sewage, litter	Tidal cycles
Minutes		Wave and currents
Seconds		Sediment grain movement

what we observe today will always be the same. An objective way to approach the problem is to regard areas of accumulated sediment (i.e. sand dunes, salt marshes, beaches) as a sediment store, a store which is not permanent, but which represents sediment that is currently surplus to the sea's requirements and that will, one day, be moved by waves and tides and subsequently deposited elsewhere. If we take this viewpoint, whenever we visit a coast, geomorphic features, such as sand dunes, salt marshes, shingle ridge, beaches, and even eroding cliffs, are seen as areas of stored sediment waiting to be eroded by the sea when conditions change. A change in conditions can occur naturally, although the real significance lies in the fact that such

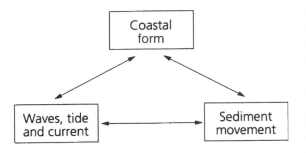

FIGURE 1.4 Simplified process diagram to show the interaction of main components of coastal processes

changes are often triggered by the activities of humans at the coast, such as the building of sea defences, the interrupting of longshore sediment transport, etc.

If we are to undertake a study of coastline management and the processes which interact along a coastline, we need to know something about how the coastline functions. If you were to study rivers, you could argue that your fundamental unit of study is the drainage basin. Similarly, if you were to study lakes, your unit of study would be the lake and its catchment. In the case of the coast, the units of study are less clearly defined and obvious. You could argue that the coastline of a particular country should be your unit of study. In some cases, it could be argued that a country has two unconnected coasts, such as the Pacific and Atlantic coasts of the USA, or the Mediterranean and Atlantic/English Channel coast of France. Even this subdivision would be unmanageable, however, and unnecessary because even on continuous stretches of coast, such as that of the UK, there is a discrete structure to coastal processes.

Since the early 1970s, coastal geomorphologists have researched and developed a structure for the coastline, by discovering that coarse sediment movement is largely restricted into specific areas, and only on extreme conditions does it move outside these areas. On the largest scale, work by the Hydraulics Research Institute has demonstrated how the coastline of England and Wales can be split into ten units, termed *coastal cells*. These cells can be used to form the basis of UK coastal management, and are, generally speaking, discrete units which function in isolation from each other. More recent work published by the Joint Nature Conservation Committee (JNCC) has extended this boundary definition to Scotland, and also undertaken some fine-tuning of some of the cell boundaries (Figure 1.5).

The concept of these cells is important and also has great significance for coastal management, although because sediment can move between cells under certain conditions, and because individual cell boundaries are not fixed, actually drawing lines on a map is not straightforward. A cell boundary will occur when longshore transport is zero. Often this occurs at headlands or across estuaries. However, under extreme wave conditions, even the largest headland can be bypassed. A second difficulty in defining cell boundaries arises from the consideration of what is longshore sediment transfer. Transfer of fine silts and clays, typically as suspended sediment, relies on tidal and water movements, and so is not as good for discerning cell boundaries. Coarser sediments, such as coarse silts and sands, need greater energy to become mobile, and so will be more restricted in their movement, and more likely to cease at areas where currents slacken. While this is not too much of a problem, when considering the concepts of coastal zone management, which is often based around the cell structure, it is important to remember that transfer of fine material will still occur if currents are too slack to move coarser sediments, and so any

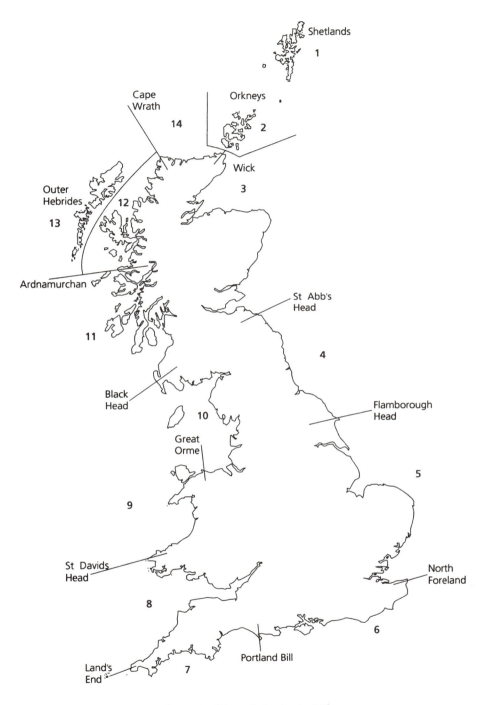

FIGURE 1.5 Delineation of coastal cell boundaries for the UK

management of sediment budgets must take this into account. That is, fine sediment can still be removed from a beach, even if longshore transport of sand is effectively zero. Despite these problems, movement of coarse sediment is largely confined within coastal cells, only transferring to adjacent cells during periods of high energy, such as under storm conditions.

Sediment input, movement, and output, whether of coarse or fine sediment, and by whatever process, are collectively referred to as the sediment budget. This budget may be determined, with adequate research, and used as the basis of coastal management strategies. As we will see later, it is vital that such considerations are taken into account in any management strategies, as to ignore sediment movement in all its forms is a serious error with repercussions for the potential success of coastal defences and for coastal stability.

As a result of this coastal cell development, and ignoring the slight discrepancies in some of the boundaries, we are able to treat the coastline in a similar fashion to a river catchment with a section of coast having distinct boundaries. The boundary discrepancies occur owing to the fact that, as we have already seen, coastal processes alter and change, and therefore cell boundaries may vary within certain definable limits. Within the limits of these large-scale cells, there are contained smaller cells, known as sub-cells. These sub-cells also function as individual process units but largely relate to nearshore sediment movement, such as processes within a bay – for example, Morecambe Bay (a sub-cell) in north-west England – or between two headlands or estuaries.

Even at this sub-cell level, however, coastlines remain areas of high complexity where sediment is eroded, transported, and deposited; where waste material is discarded; where resources are exploited. The net result of all of this activity is that the coastal environment is subject to a complex interplay between natural and anthropogenic processes, which can occur either in harmony or in opposition.

Estuaries have long been of great importance to humans, whether as places for the development of settlements and industry, for sources of water, or for communication. As a result, they have traditionally been the recipients of much attention, and have been highly modified, both directly by land claim, and indirectly as a site for waste disposal.

Estuaries are extremely productive habitats in the ecological sense. It has been claimed (McLusky, 1989) that estuaries may even be regarded as the most productive ecosystems in the world, albeit with much of this productivity going on against a background of human interference. Estuaries have often been regarded as natural waste-disposal systems and as a method of discharging waste to the large oceanic sink, where it will be dispersed. This, however, is not always the case, with many of the pollutants staying in the estuary (Chapter 4). As a result, many estuaries have experienced the accumulation of pollutants, both within the organisms which live there and within the sediments themselves (Chapter 4).

As well as considering what goes on at the coast or within estuaries, it is also important to remember that the two environments can be affected by other processes which operate beyond their boundaries. Estuaries represent the place where rivers meet the sea, and many pollutants, for example, can be discharged into the rivers well upstream from the estuary, at any point within the catchment. Estuaries such as that of the Severn in south-west England, for example, may contain within their catchment large industrial areas and large cities; the Severn drains large parts of the Midlands and southern Wales. Similarly, large land clearances or dam constructions can dramatically increase or decrease, respectively, the amount of sediment reaching the estuary or coastline. Clearly, when a sediment budget is being considered, a significant alteration in sediment supply can produce major changes in the sediment regime of a stretch of coastline or an estuary.

◆   ◆   ◆

Historically, many of the problems with human intervention have arisen from the ethos that we have to protect and enshrine the coast as we find it, or – put another way – we must never allow the sea to take land from us. As we shall see in this book, this is not always the most sensible policy to adopt, and in the long run we may well be better off allowing part of the coast to be taken by the sea and, as a result of this sacrifice, thereby increasing the protection of the coast elsewhere by natural processes, without resorting to expensive forms of defence. In effect, by piecemeal defence, we merely cause problems elsewhere, in some cases to an extent far worse than the problem which the original defence works set out to cure. It is important to realise, therefore, that putting a sea wall against an eroding cliff is not necessarily the best approach. Certainly this method may well protect the cliff, but elsewhere the fact that the cliff is no longer eroding means that sediment supply to adjacent beaches will be reduced (see Chapter 3). As beaches represent areas where sediment is constantly added and removed (the net balance of these two processes will determine whether the beach is accreting or eroding), the removal of sediment input from a newly defended cliff may alter this balance between addition and removal, with the result that a beach which once received more sediment than it lost may well subsequently receive less sediment, causing erosion to become the dominant of the two processes.

The distinction of coastal type

Coasts may be studied in several ways, depending on the way in which the observer classifies them. Some observers regard coasts as 'erosional' or 'depositional' (see Davis, 1996), while others prefer a distinction on the basis of sediment type, such as coarse clastic coasts (shingle and sand) and muddy coasts (see King, 1959). Others describe coasts as being emergent or submergent (Johnson, 1919; Valentin, 1952), or classify them according to tectonic setting (Inman and Nordström, 1971). For this book, we shall adopt a more process-based classification, because from our point of view it is the human intervention in the natural processes which brings about many of the problems which will be discussed. Hence, coasts will be divided into three main groups: wave dominated, tide dominated, and wind dominated, depending on the dominant process responsible in their formation (see Hansom, 1988). Each dominant process is associated with distinct depositional environments (Table 1.2).

TABLE 1.2   The contrasting forms of agent dominance at the coast and their resulting landforms

Wave dominated	Tide dominated	Wind dominated
Shore platforms	Mudflats	Sand dunes
Cliffs	Sandflats	
Beaches	Salt marshes	
Spits, tombolos	Mangroves	
Deltas	Deltas	

Energy

High ⟶ *Low*

It is important to bear in mind, however, that these terms refer to dominant processes, and as such do not exclude all others. For example, a tide-dominated coast will, morphologically, reflect tidal processes, but will still experience waves and wind. The primary difference between the three coastal types is based on energy (Hansom, 1988). Wave-dominated coasts are high-energy environments, and as a result contain features such as cliffs (eroded by high-energy waves), and sand and shingle beaches. Fine sediments do not get deposited because the energy levels are too high to allow them to settle out. Tide-dominated coasts are lower-energy environments and so reflect this in mud and silt deposition in the form of mudflats, salt marshes, and mangroves. Wind-dominated coasts are slightly different in that wind is very much a modifier of the other coastal types, particularly wave-dominated. The wind is responsible for the transfer of sediment from the intertidal zone into the supratidal zone, causing the formation of features such as dunes. The amount of energy available is also reflected in the erosive capacity and the sediment transport potential of a particular environment. These processes are summarised in Figure 1.6.

Human impacts on coasts and estuaries

While it is true to say that the coastal environments listed in Table 1.2 owe their origins to natural processes, many of them have been subsequently influenced by the action of humans, albeit to varying degrees of intensity. It is also true to say, however, that in some cases the natural environment has also had an important influence on humans.

Humans have a long history of coastal and estuarine exploitation. The impact of this exploitation will vary according to the type, and also the sensitivity, of the environment. For example, a dune system will be more sensitive to trampling than a shingle beach, so that the same act of walking will cause different degrees of damage. Table 1.3 lists some of the main uses of coastal environments, and also details how these uses may impact on the environments themselves.

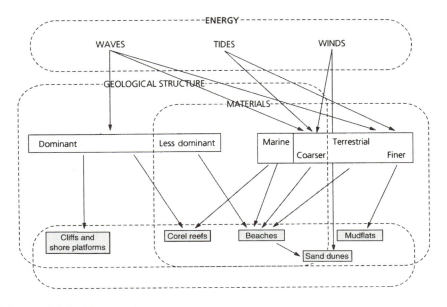

FIGURE 1.6 The coastal system
Source: Hansom, 1988

Human activity at the coast can be divided into seven key areas. These areas are introduced briefly here, but discussed in greater detail in following chapters.

Building development and land claim

The pressure for developing areas adjacent to the coast is generally great, whether for ports and harbours, large industrial plants, power generation, residential development, or leisure. Demands for land can be met in one of several ways:

1   developing greenfield sites;
2   redeveloping previously used sites;
3   making new land for development by claiming intertidal areas.

In the case of land claim, the need for new land is often weighed against other factors, such as waste disposal, where wastes are used to build up the level of the intertidal area as a form of landfill.

In the past, piecemeal land claim and poor understanding as to who is the overall authority in estuaries (that is, where does land law stop and sea law start?) has led to damage to various habitats, resulting in interference with natural processes and the subsequent need to protect unsuitable development by means of defences. This problem is often aggravated by the fact that many coastlines fall within the jurisdiction

TABLE 1.3   The potential human impacts on coastal environments, and their significance with respect to change

Environment	Human impacts[a]	Sensitivity to change
Shore platforms[b]	Collecting	Low
	Education	Low
	Walking	Low
Cliffs	Cliff-top development	High
	Climbing	Low
	Education	Low
	Coastal defences	High
Beaches	Recreation/tourism	High
(sand & shingle)	Coastal development	High
	Coastal defences	High
	Sand/shingle removal	High
	Trampling	Low
	National defence	Moderate
	Off-road vehicles	Locally high
Mudflats/sandflats	Land claim	High
	Bait digging	Generally low
	Industry/port development	High
	Dredging	High
	Coastal defences	High
	Pollution	High
	Conservation	Low
	National defence	Moderate to low
	Estuarine/tidal barrages	High
	Marina development	High
Salt marshes	Land claim	High
	Grazing/harvesting	Moderate to low
	Recreation	Low
	Port/harbour development	High
	Dredging	Moderate to high
	Conservation	Low
	National defence	Moderate to low
	Coastal defence	Low
	Pollution	Low
	Estuarine barrages	High
	Marina development	High
	Off-road vehicles	Locally high
Mangroves	Land claim	High
	Logging	High
	Aquaculture	High
	Agriculture	High
	Pollution	High

TABLE 1.3   continued

Environment	Human impacts[a]	Sensitivity to change
Dunes	Land claim	High
	Recreation	Locally high
	Trampling	High
	Sand extraction	High
	National defence	High
	Coastal defence	High
	Conservation	Low
Offshore	Waste discharge	Potentially high
	Fishing	Potentially high
	Coral reefs	High
	Extraction	Moderate to high
	Waste dumping	Locally high

Notes

[a] Denotes impact on *natural* habitat, and therefore includes habitat modification.

[b] Excluding beaches.

of several local authorities; thus planning policies within any one estuary can vary. The Dee estuary in the UK, for example, falls within the control of two county councils, six district councils, and two countries (see Chapter 7), and so what may be suitable development in one authority area may increase problems in an adjacent area.

In order to maintain the sustainable development ethic and reduce environmental impacts, the way forward is to seek alternatives to developing new areas and thus remove the need to claim intertidal areas. One solution, stopping development altogether, is not practical. The ideal is to undertake development in such a way that there is only minimal conflict between different user groups and human/natural processes. The two main features which need to be discussed in development are:

1   the need to consider wider economic, social and environmental implications of decisions and actions;

2   the need to take a long-term view of any action proposed.

In reality, the use of management plans (Chapter 7) is considered the most important tool for identifying where development may occur, and to reconcile sustainable use with other interested parties. In addition, any new development will be allowed on new sites only if it is the sole possibility, and no alternative sites can be found. By adopting such an approach, the need for new defences can be minimised, thus keeping as much of the coast as possible free for natural sediment cycling.

Navigation, dredging, and aggregate extraction

The dramatic increase in leisure boating has created a great demand for the provision of sheltered moorings and marinas. Where these have been developed within existing facilities, the threat to the environment has been slight. Problems arise during the construction of new facilities, where land claim and dredging need to be undertaken. Additional problems arise when there are demands for additional facilities to supplement the marina development, such as infrastructure, servicing, and housing. The impact of marina development can be gauged from the fact that there is not one major estuary in the UK without plans for some form of marina construction, or where development has already occurred (Pye and French, 1993a).

Problems relating to navigation and access for shipping become of crucial importance in considering shipping to ports and harbours. As containerised ships, tankers, and ferries increase in size, the depth of water needed to allow them access to berths increases. In order to remain viable as ports, harbour areas often need to undertake extensive dredging programmes to deepen the approach channels.

Dredging can be classified into two types: maintenance dredging, where existing channels are maintained at a specified depth; and capital dredging, involving new works to deepen existing channels, or to make new ones. The problems associated with making artificial channels can have severe implications on the coastal and estuarine environment. A dredged channel can completely modify tidal flow, ebb and flood tide directions, the tidal prism, and the intertidal profile, which will naturally respond to changes in channel depth by attempting to level out; that is, with a deepening channel, the intertidal profile will generally become steeper (Pye and French, 1993b). Thus intertidal and subtidal hydrology can be significantly altered or modified. These problems will be explored further in Chapter 3.

In addition to the morphological changes, the actual process of dredging will cause the suspension of large amounts of sediment, which will make the water body less clear, and can have significant implications for benthos (Bohlen *et.al.*, 1979). A second problem is what to do with the material which has been dredged. If this is suitably pollutant free, it can be utilised elsewhere to build up beaches, habitat recreation, or offshore bars for coastal defence purposes. As an example, the spoil from recent capital dredging works at Harwich Harbour, Suffolk, has been used in Hamford Water to recharge beaches, and construct an offshore bank to reduce the problems of wave erosion within the salt marsh area.

The extraction of gravels both from beaches and offshore has similar effects to dredging, with the added problem that offshore bars can be removed, thus exposing the coastline to increased wave attack. Well-planned schemes should not be subject to such problems, but in some cases disastrous results have arisen as a result of gravel extraction. In the 1890s, dredging occurred within Start Bay in Devon, south-west England, to obtain gravel for the construction of Plymouth Harbour. It was estimated that material removed would be replaced by natural processes and no net environmental change would occur. In effect, no replacement occurred and the increased wave attack on the shoreline eroded the beach and the cliffs. As a result, the village of Hallsands, located behind the shingle beach, was lost to the sea (Job, 1993).

Energy generation/barrage construction

Energy generation is of vital importance to human well-being. In the past, generation has generally been via coal, gas, oil, or nuclear power stations, sites on estuaries being generally favoured owing to the sheltered location and plentiful supply of cooling-water. In their own right, these sites demand a lot of land, as well as berths for the supply of raw materials.

The more contemporary policy of renewable energy has led to a change of ideas concerning tidal, wind, and wave power. Wind energy is not an estuary problem and so need not be considered here. Wave power is achieving some interest but is still very much in its infancy. The largest threat comes from tidal power and the construction of barrages, both for power generation and for amenity usage.

Any barrage development will have a dramatic influence on the estuarine regime, especially considering that the most suitable estuaries have the greatest tidal range, and, therefore, the largest area of intertidal flats, important for bird feeding habitats. If a barrage is built, the intertidal area is dramatically reduced, affecting bird feeding areas and changing the whole hydrodynamic regime of the system. This change takes the form of alterations in the flushing of pollutants, sediment dispersion patterns, silting, and port access. Additionally, a lake, commonly the result of a barrage, is a tremendous attraction to the leisure industry for sailing, etc., and thus secondary development is often great.

Barrages are extremely detrimental to the estuarine habitat, and completely destroy the sustainable development ethic. Before barrages are built for power generation, two main points need to be taken into account (Pickering and Owen, 1994; Middleton, 1995):

1   The return on capital investment is minimal.
2   The amount of power generated by destroying all the UK's main estuaries would provide around 10 per cent of UK energy needs. Effective use of existing energy supplies, however, would save around 40–57 per cent of existing generated electricity. Hence other, more suitable, methods exist for increased energy provision without the need to destroy such estuarine areas.

When considering the generation of power by any means, there are several problems which need to be tackled:

1   Offshore exploration (oil and gas) can cause pollution problems and development pressures on land and in estuaries.
2   Power generation in an estuary can lead to the input of large quantities of warm water into a partially land-locked body of water, causing localised warming. This can affect fish populations and also cause novel problems, such as that experienced in Shoreham Harbour, at the mouth of the river Adur, Sussex. Here, the warm water outflow from the power station into the harbour area led to sufficient warming that coral polyps, brought in on the bottom of banana boats and the like, started to grow, with the result that a coral reef

began to develop in the harbour area (Shoreham Port Authority, personal communication).

3   The import of raw materials can cause pollution problems.

The development of power plants often occurs at the coast, for reasons already stated. Clearly, these are of high land value and need to be protected for their operating life. In addition, however, nuclear power stations pose a further pollution threat, in that once at the end of their operational life, they are still highly radio-active. For this reason, their protection from erosion must be assured for two to three hundred years, at least, after they have finished generating power. Normally defences are constructed with anything from fifty to a hundred years of operating life. Hence, the location of nuclear power stations on the coast gives the coastal defence authorities the problems of a commitment to defend under whatever conditions of erosion, sea level rise, storminess, etc. happen to operate in over two hundred years' time.

Fisheries

Many commercial fisheries are situated near to the coast or in estuaries because of the relatively high productivity and sheltered nature when compared to the open sea. There are already tight controls on management practices, but outside activities can have a dramatic influence on fish farming concerns.

Bad management of fish populations can alter the ecological balance of the system, such as by allowing the presence of too many predators or too many prey species, which will inevitably cause a knock-on effect and influence the food chain. In addition, altering the fish population can also alter the bird population, as hunting birds are specific about what type of fish they eat.

Outside activities, such as dredging and gravel extraction, can destroy fisheries by increasing the amount of sediment in the water body. Also, marsh management techniques, such as spraying to reduce floral growth, can lead to increased susceptibility to erosion of mud flats, and thus to increased sediment loads.

The presence of local fisheries will have important implications for just what forms of defence are tolerable. Clearly, anything which allows increased sedimentation could destroy fish beds. This problem has occurred along the frontage at Morecambe in Morecambe Bay, where the installation of the new fish-tail groyne defences has proved so effective in trapping sediment that deposition has occurred in areas of once viable shellfish beds, decimating the local shellfish population and ruining the livelihoods of some local people.

In effect, the presence of fisheries means that actions by the coastal defence authority have to be particularly sympathetic to coastal users, in that large job losses as well as ecological imbalances can occur if the wrong, or inappropriate, methods are used.

Agriculture

The association of agriculture and estuaries goes back a long way. Historically, one of the main reasons for land claim has been for agriculture, and also for the diverse pasture which marshes offer for grazing. The knock-on effect of this is that there are large areas of farmland in many countries which, without adequate coastal protection, would be inundated by the sea.

Changes in agricultural practices now mean that little land is being claimed for this purpose. In Europe, land put into set-aside by European Commission policy is often targeted for use in managed realignment schemes, a form of soft engineering in which land is allowed to flood at each high tide to generate new salt marsh. Similar policies occur in other countries. Titus (1991) details a case in the USA where similar actions could result in the sea being allowed to flood an area of land equal in size to the state of Massachusetts. As a result of such practices, low-grade agricultural land, or other such low-value land, is being turned over to new uses, primarily in the potential abandonment of it to the sea. This activity is now forming policy for many government agencies, with farmers being paid to allow land to revert to intertidal habitats. Such schemes are being employed particularly in areas where coastal squeeze is acute, and where new intertidal habitats are needed for defence purposes (see Chapter 3).

Problems of agricultural run-off can also bring about nutrient enrichment and increased sediment loads, leading to eutrophication and silting problems. The presence of marshes helps to lock up nutrients, as does taking land out of cultivation where the developing sward will lock up nutrients and also reduce sediment run-off.

Water quality

Any organism that lives in water, or depends on the many activities that occur upon it, relies on water quality. The definition of 'clean' water will vary with any one particular interest; that is, certain pollutants may not affect some organisms but will others. Generally, however, the concept of industrial discharges is difficult to reconcile with the concept of sustainable development.

As well as there being pollution in the water body, heavy metals, for example, also become adsorbed onto sediment particles and so become stored along with deposited sediments within salt marshes and mudflats (see Chapter 4). Hence, estuaries with highly industrialised catchments, such as the Tees and Severn in the UK, contain extensive stores of metals in their sediments (Alexander *et al.*, 1936; Taylor, 1979; French, 1993a, b, 1996a, b). From a coastal defence point of view, stores of pollutants in sediments can become remobilised if those sediments start to erode. This situation may occur where new defences are constructed nearby, or where the sediment budget is modified in some form. Such processes can lead to increased wave attack, or increased erosion, which serves to reintroduce pollution into the estuary system. Such a source of pollution can be significant in some cases and, whereas industry has emission controls placed on it to restrict the amount of pollutants discharged, salt marshes do not, with the result that in some areas large quantities of contaminants can re-enter the aquatic system (see Chapter 4).

Sports and recreation

Perhaps the greatest impact on coasts and estuaries is from the leisure industry. With the expansion of leisure time and the increase in demand for facilities, the demands on coasts and estuaries are ones which need to be carefully managed in order to protect sites for future uses, and the coexistence of other interests.

Some coasts experience greater impact than others. Tidally dominated coasts, for example, do not generally appeal for beach holidays, but more to bird watchers, walkers, etc. For this reason, they tend to avoid the large-scale tourist development which can mean serious environmental degradation. However, these low-energy coasts do tend to attract other interested parties, such as those already discussed, and so tend to be exploited in other ways (see Chapters 3 and 4).

Wave- and wind-dominated coasts are the most popular for tourism. Sand and shingle beaches make suitable sunbathing sites, and so these areas are often heavily developed and heavily used, and, as a result, often raise their own set of defence and management issues because defence is not just a case of stopping erosion or flooding, but has to occur in such a way that access and aesthetic qualities are retained.

The sorts of problems which often occur in resorts are:

◆ Marinas are developed.
◆ Modification of beach profiles to provide gentle slopes for sunbathing can often destroy the natural profile and, hence, modify the effect of waves and tidal processes on the beach environment.
◆ Wind- and wave-dominated coasts can be backed by extensive sand dune fields, which are generally very fragile environments. These areas are often appealing to holiday-makers for exploration, sunbathing, picnics, parties, trail biking, etc. For this reason they receive a lot of abuse (see Chapter 5).

◆   ◆   ◆

So far we have considered seven prime uses of estuaries and coasts, and I have introduced the types of activity which occur and the problems caused. These issues will be returned to in detail later and will form the basis of this book. There is one additional coastal activity which links all of the above. Rather than being considered as a coastal 'use', however, it is very much a coastal response. This response may be to natural processes, but it is more frequently a response to activities associated with the seven uses mentioned above. This activity is coastal defence – that is, the construction of hard or soft defence measures to protect the land – and is perhaps the most provocative activity in relation to the coastline today. The traditionally held view is that any development along the coastline claims a right to some form of sea defence. In estuaries, some of the problems associated with the open coast are minimal, but that of erosion is still very much in evidence. In total, around 25 per cent of the coastline of England and Wales is developed in some way (Pye and French, 1993a). Areas of intense development are referred to as being land of high value, and are generally defended and protected. Areas of farmland, however, much of which was previously claimed from the intertidal area in the first place, are generally considered of lower value, and so may not be defended once existing defences have reached the end of their life span, or are breached by storm action. This policy of not protecting farmland or other 'low'-value land leads to inevitable conflict where isolated houses and farms are allowed to fall into the sea because coasts are left to go on eroding without any protection measures being taken.

In the south-east of England where relative sea level is rising to the greatest extent, partly as a response to global warming, to tectonic downwarping, and to the isostatic adjustment of Britain following the last glaciation, coastal defences (typically hard engineering structures) have been constructed on a piecemeal approach, with the result that one local authority has produced greater problems for its neighbours by cutting off the supply of sediment which originally went towards neighbouring beaches. In a natural system, sea level rise will be compensated for by a natural land-ward movement of habitats. However, where there are hard sea defences such as sea walls, this cannot happen and such habitats are lost by a process of coastal squeeze (see Chapter 3).

The notion of sustainable development along the coast – that is, the maintenance of habitats – is noted in the ethics of soft engineering methods of coastal protection. Instead of the building of large concrete structures or boulder revetments (hard defences), soft engineering techniques, such as beach feeding, marsh regeneration, offshore breakwaters, or managed realignment, are the preferred options, where the natural line of defence is promoted by techniques sympathetic to ecosystem and coastal processes. In other words, if the conditions for marsh growth or sedimentation are artificially managed, the system will naturally look after itself. These processes are described and discussed in Chapter 3. This idea of soft as against hard engineering underlies much of our discussions so far. Areas of intensive land usage and high land value need to be defended and so some intervention is necessary. Also, hard engineer-ing methods are less favoured than soft methods, because soft methods allow the coast to continue to operate in as natural a way as possible. Hence, the impact of human interference from a defence point of view is considerably less when soft methods are used than when hard methods are used, and in many circumstances the eventual outcome is the same, if not more favourable in the case of soft methods.

◆   ◆   ◆

Before we can consider these aspects, however, there is one key area of importance which needs to be covered first. This concerns exactly how the problems which have been mentioned arise. First, therefore, it is necessary to consider these natural coastal processes. The next chapter will introduce the types of natural processes which occur along wave-, tide-, and wind-dominated coasts, and also the ways in which human intervention can cause these processes to change and, as a result, produce adverse environmental impacts. Having considered these processes, we will be in a position to achieve a greater understanding of how the activities detailed earlier, and covered in more detail in Chapters 3 to 6, actually cause problems.

Summary of Chapter 1

This chapter has introduced the background issues needed for consideration of the following chapters. The issues which are briefly introduced form the backbone of the human impact problem. In addition, the chapter also introduces the problem of coastal erosion as a global problem. The clear message of the work cited, and also from other work indicated in the references section, is that while natural processes of isostatic land movements and the quasi-natural aspects of sea level rise are unquestionably important, the overriding issue comes down to the human factor. The prime issue is the role humans play in controlling the amount of sediment available to the coast. Whether by preventing cliffs from eroding, stopping longshore movement, or preventing sediment from reaching the coast in the first place, the outcome is the same. Coasts will only be stable or accrete if there is sufficient sediment available. Stable coastlines are not coasts which do not interact with the sea, they are coasts on which sediment accretion is occurring at the same rate as sediment erosion, and hence they still need a supply of sediment to maintain their form. As soon as any human action reduces the amount of sediment reaching a particular coast, the sea will start to erode its sediment stores; that is, those areas of sediment which the sea has previously deposited. These stores are perhaps better known under their normal names: sand dunes, spits, beaches, marshes, mudflats, etc.

This really forms the basis of all the issues which we will discuss. Whatever impact we are considering, it will always come down to the fact that the coast erodes because it does not have enough sediment to sustain itself. All that changes in each case is that the method which is employed to deprive the coast of sediment changes.

There have been many important concepts introduced in this first chapter. The reader should leave it with an understanding of the following points:

◆   the general problem of coastal erosion from a global perspective, and the causes for it;
◆   an appreciation of how coasts behave, and the concept of coastal cells;
◆   the issues which will be tackled in the rest of this book;
◆   the underlying basis of all coastal erosion problems: the control and quantity of sediment.

The physical
regime of coasts
and estuaries

Introduction

The previous chapter introduced the basis of material covered in this book. Before we look in detail at the various impacts of humans in coasts and estuaries, it is important to acquire an overview of how these systems operate, because without this, it is not possible fully to appreciate how human activities can cause the impacts which they do. As was said in the last chapter, it is not natural processes which produce the problems experienced along the coast, but the human activities which interfere with these processes, and also the development in areas affected by them.

The problems which we can observe at the coast today all arise out of the fact that humans have, in some form, affected the natural processes. Because of this, it is necessary to look at the coast from a process point of view. Table 1.2 introduced the three-stage classification of coastlines based on the three main process agents – waves, tides, and wind – and indicated the various coastal landforms associated with each. We will now consider each type of coastline in detail with respect to the dominant process. From this, an understanding will develop as to how delicate this equilibrium is, and how the actions described in the following chapters can cause the impacts which they do.

Wave-dominated coasts

As the name suggests, the most important effect on wave-dominated coasts is that of waves, although wind and tides can still play a part. Such coasts typically are exposed ones, open to the sea and with little shelter from wave activity.

The degree of wave activity will govern the types of sediment accumulating, but a basic relationship makes it easy to understand the work that waves do:

$$\text{Wave energy} = \text{coastal erosion} + \text{sediment transport}$$

If there is little sediment to transport, then the bulk of the wave's energy will be focused on erosion, although other factors, such as the angle of wave approach, are involved. If there is a lot of sediment, then the coast will be largely erosion free. This explains the cyclical nature of much of the coastal erosion which occurs. Cliff erosion, as an example, is always quoted an average rate of x m yr^{-1}. However, in reality, cliffs tend to erode at regular intervals. A cliff fall may happen one winter, but then no more falls will occur for a period of years. The reasons for this can be traced back to the relationship given above. A fall occurs when there is little beach sediment to erode, therefore the wave energy is focused on erosion of the cliff. When a fall occurs, a large volume of sediment lies at the base of the cliff, and most of the wave energy is employed in moving it. When this has been done, energy will, once again, be used for cliff erosion. This also highlights another important issue, that of the importance of sediment in the protection of coasts from erosion, an issue to which we shall return in Chapter 3.

The distinction between erosion and sediment movement is not necessarily clear-cut because it can be argued that the moving of sediment off a beach is erosion. However, a constant theme to which we will be returning is that sediment on the coast does not represent a permanent landform, but a temporary store of sediment. Naturally, if wave activity is sufficiently great, than all sediment will be removed from a beach, creating a situation of no sediment accumulation, and a bare rock surface. In these cases we are left with a feature known as a shore platform. In some literature, shore platforms are referred to as wave-cut platforms, although because this implies a single formative process (namely, the action of waves), the term is now commonly replaced by the less process-suggestive term 'shore platform'. Conversely, if wave activity is slight, the energy will not be sufficient to move sediment off the beach, but will only be able to refashion the beach profile (that is, move sediment around). In these cases, landforms such as beaches will result. As the energy of the waves increases, so the waves have greater ability to move sediment around. In addition, the action of waves starts to produce current activity, and thus longshore movement starts to occur.

On wave-dominated coasts, energy is high and therefore fine sediments are generally absent. Depositional areas generally include sands and shingles; that is, beach forms. Longshore currents can refashion these deposits into secondary features, such as spits and tombolos. Erosional aspects include cliff formation and mainte-nance, as well as the remobilisation of beach material. Typically, such erosion is regarded as a loss of land and, therefore, something to be avoided and prevented. If, however, the same problem is looked at from a process and coastal management point of view, then erosion is seen not just as the loss of land, but also, and perhaps more importantly, as the supply of sediment to the coastal zone. This sediment is critical to the build-up of beaches and other coastal features elsewhere. From a human impact perspective, the key problems which arise along wave-dominated coasts stem from the interference with these areas of erosion (cliff protection, beach protection), and also the interruption of longshore sediment movement.

The distinction of whether a wave-dominated coast is erosional or accretional lies partly in the form of the wave itself, and partly in the form of the beach profile (Figure 2.1). It can be seen from Figure 2.1 that wave energy is the critical point, and the resulting processes are governed by the amount of surplus energy available. Broadly speaking, waves can be classified as either constructive or destructive. Constructive waves typically input more sediment than they remove, thus giving a situation of net sediment accretion. Destructive waves remove more sediment than they deliver, resulting in net sediment removal. Coasts often experience both wave types, often on a seasonal basis. This explains why beaches change shape from summer to winter, with material being drawn down in the winter to produce a flatter beach profile which is more efficient at dissipating wave energy. While it is not the

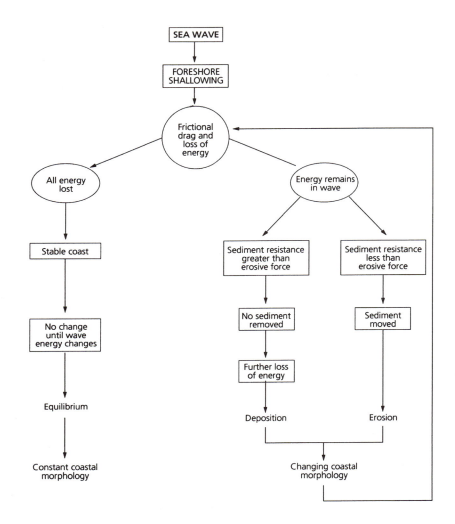

FIGURE 2.1 Environmental controls of sediment deposition and erosion in a wave-dominated beach system

intention of this book to look in detail at wave movement and energy transfer, coastal management does require some understanding of wave energy transfer. Waves are generated when the surface of the sea is disturbed, such as by wind, gravity (tides), or seismic activity. There are four main categories of wave:

1   capillary waves – these are small waves with frequencies of < 0.1 seconds and are of little significance for coastal processes;
2   gravity waves, or wind-generated waves, which tend to have frequencies

between 1 second and 15–20 seconds; these are perhaps the most important waves for coastal processes;

3   infra-gravity waves with frequencies in the order of minutes occur as the result of the interaction between two sets of gravity waves;

4   long-period waves caused by tidal forces; that is, tides, which can be looked at as single long-period waves of *c.* 12 hours' duration with the wave crest equating to high tide and the wave trough equating to low tide.

When one is considering how waves impact on the coastline, it is important to be aware of several facts. First, it is the wave form that moves, not the water. Water molecules will move only in a circular or elliptical motion (Figure 2.2). In open water, water movement at the water surface is circular. Towards the base, the proximity of the bed and increasing stresses cause the circular orbits to become distorted and more elliptical. As the wave moves closer stil to the shore, this distortion increases until a point where the ellipses fail to close and water movement is then initiated forward and backward, hence waves start to break. This point is known as the surf line.

Second, the depth to which these orbits penetrate is also important. In deep water, this depth, known as the wave depth and approximately equal to half the wave length (Carter, 1988; Hansom, 1988; see Figure 2.2), does not reach the sea bed, and so the waves do not disturb the bottom sediment. This sediment is largely immobile in that it is not available for transfer to beaches unless the wave base deepens owing to more energetic wave activity. Closer inshore, however, as water shallows, the wave base is deeper than the bed, and so the ellipses caused by water movement can interfere with bottom sediments, causing sand to enter suspension, and larger sediment grains to be moved around. This point typically occurs when the water depth (synonymous with wave base now that we are inshore) equals a quarter of the wave length (Carter, 1988). As the water becomes shallower still, so this process of sediment movement and redistribution becomes greater. When this process occurs, we start getting into the realms of sediment movement and transfer. The point at which wave base meets sediment interface is not a constant line along the shore, but will vary according to wave conditions. In stronger wave climate, where waves get bigger and wave base deepens, more sediment falls within this zone, while in calm areas, more of the sediment is left undisturbed by wave activity.

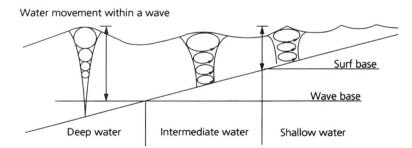

Water movement within a wave

Surf base

Wave base

Deep water   Intermediate water   Shallow water

FIGURE 2.2   Effects of foreshore shallowing on the wave form
Source: Hansom, 1988

This basic principle of wave motion and sediment movement will occur with just a change in magnitude as wave energy increases. During a nice quiet summer's day, wave action is minimal and so wave base is very shallow, hence little of the bed is actually affected by it. Under storm conditions, where wave height can be of the order of metres and wave length greatly increased, wave base is obviously going to become deeper owing the relationship between wave base and wave length. Because of this increase, more of the bed will be influenced. As wave energy increases and more of the bed is influenced by wave action, so more sediment can be moved around. As a result, storms will cause a greater change in beach form because of the increased wave length and increased wave base, both of which are related to the increased wave energy created by storm conditions.

As waves move onshore, the sediment surface grades upwards and water depth shallows. This process of shallowing also accounts for another feature often observed along coasts. When a wave breaks and nears the shore, it gets higher. This effect is partly due to the physical laws of the conservation of energy, but also due to friction with the bottom sediment. The increased friction is greater at the bottom of the wave than at the surface, and so the wave base receives more frictional drag than the surface waters, with the result that the surface waters travel faster. Owing to this variation in speed, the surface waters 'overtake' the water at the sediment surface, forming the characteristic breaking waves.

It is possible to see, therefore, that given this relationship, the shape of the intertidal profile will be important in the consideration of the effectiveness of wave activity on the shoreline. As soon as a wave makes contact with the bottom sediment, it starts to lose energy through friction. As a result, the longer that a wave experiences this friction, the more energy it loses. It can be seen, therefore, that a foreshore which is wide and shallow will cause a wave to lose more energy through friction than one which is short and steep. From a coastal defence point of view, we can immediately see one way in which wave activity can be reduced at a shoreline, and any activity which causes a beach profile to steepen will result in that coastline receiving an increase in wave energy.

We have seen that the water depth controls wave height under the physical laws of the conservation of energy. This occurs primarily as a result of the fact that the wave slows when it enters shallower water. This shallowing of the foreshore is also responsible for another important aspect of wave action. If the wave form is moving perfectly parallel to the shore, then this decrease in speed due to friction with the bottom sediments will be uniform. In reality, however, most waves approach the coastline at oblique angles; thus parts of the wave will be in slightly deeper water than other parts, meaning that the portion of wave in deeper water will move faster than the portion in shallower water. The resultant differential in wave speed means that the wave becomes refracted. On a coastline of bays and headlands, water depths approaching the bays are generally deeper, while those approaching the headlands are shallower; thus waves are refracted towards the shallower water, concentrating wave activity onto the headland areas (Figure 2.3). The initial implication of this is that there are greater erosive forces directed towards the headlands. In addition, however, there are other, perhaps more significant, implications. After all, the headlands are

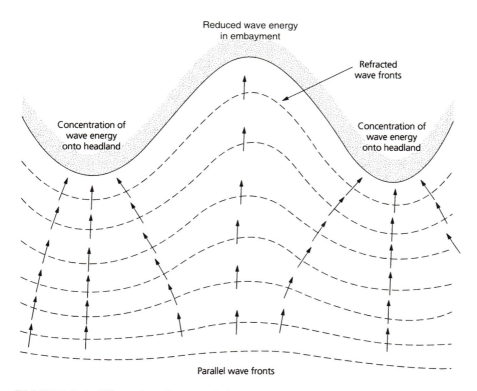

Reduced wave energy
in embayment

Refracted
wave fronts

Concentration of
wave energy
onto headland

Concentration of
wave energy
onto headland

Parallel wave fronts

FIGURE 2.3 Effects of sea floor morphology on wave front refraction, and the subsequent concentration of waves at headlands

headlands because they are more resistant to erosion than the embayments. One of these implications is that along the coast there will be a series of wave foci in which higher levels of wave activity are present than elsewhere. Thus, a series of spatial wave gradients are established which are extremely important in developing longshore currents, and in the significance of longshore sediment movement and related morphological changes to coastal formations.

It is at this point that we need to further complicate the situation. So far we have considered how energy can be lost as a wave approaches the coast, and how waves can become refracted as a result of the angle at which waves reach the coast. While these 'primary' waves lose much of their energy in this way, it is also common for a series of secondary wave forms to develop, termed 'edge' waves. Edge waves tend to remain within the surf zone and are formed as primary waves lose energy as they approach the shoreline, with some of this energy being transferred to edge waves. Although detailed explanations of these features are not warranted here (see Carter, 1988, for a detailed analysis), they do need to be considered because they are thought to exert a major influence over many coastal features. This impact is brought about by the interference of edge waves with the incoming primary (gravity) waves.

Davis (1996) explains how edge waves effectively behave like standing waves, and so can amplify the incoming gravity wave when the crests of the two wave forms coincide (according to the physical laws of wave harmonics – see Carter, 1988, or Davis, 1996). This amplification can increase the impact of waves on the shoreline, leading to greater changes in beach profile and potentially greater volumes of sediment movement.

It should also be understood that this form of wave behaviour is not restricted to cliffed coastlines. Open-beach coasts can also develop such phenomena, leading to the formation of complex beach morphologies, including bars, cusps, and washovers (Holman and Bowen, 1982).

Beach processes and human interference

Considering the high-energy regime of the wave-dominated coast, beaches may appear to be rather unlikely landforms to be found on these exposed coasts. Being piles of shingle or sand must make them susceptible to erosion and removal. In reality, the fact that they are loose material means that the waves will refashion them and reshape them, but will not necessarily remove them, unless the wave strength reaches a critical threshold. In a beach, much of the energy is dissipated within the sediment by water draining through the shingle or sand and returning to the sea by throughflow. The degree of compaction and sediment grain-size factors will control this flow. If compaction is great, for example, then little water will be able to soak into the sediment, and so much of the water will return to the sea by surface flow. As soon as this starts to happen, there is a greater tendency for sediment removal from the beach.

Beaches are also perhaps among the most heavily used of our coastal environments, and they are consequently the most heavily threatened. It has been estimated that world-wide around 70 per cent of all sand and shingle beaches are showing signs of erosion of some sort (Bird, 1985). Despite this, beaches continue to be heavily used, often being artificially maintained with respect to profile by sediment entrapment between groynes and sand import from other areas. Because of this high degree of 'artificialness', many tourist beaches need constant interference to maintain them. However, with such continuous artificial modification, they have ceased to be in natural equilibrium with the environment and so often have to be kept in place by artificial means, which has implications for the coastal cell and sediment transport patterns.

As well as being heavily used for recreation, beaches are vitally important in the role of coastal defence. Waves hitting a long, shallow beach profile will dissipate the greater proportion of their wave energy before hitting a sea wall or cliff. Hence, a healthy beach is of major importance in the coastal defence of an area, and its presence can reduce the need for large sea walls or other coastal defence measures. The slope of the beach is vitally important for effective defence; however, beach slope varies considerably with time of year, being a good example of coastal dynamism, due to variations in wave activity. During the winter months, the greater incidence of storm waves means that the beach experiences net erosion and adopts a shallower profile with material drawn down and transported further offshore. This, in effect, produces a wider, shallower beach profile which is more effective in the dissipation of wave energy. In times of lower wave strength, typically during the summer, there is a greater net movement of material from offshore onto the beach, giving the beach a swell profile. This process serves to illustrate the concept of the beach as a temporary sediment store. The material stored in the steeper summer profile is drawn down in the winter to produce a more elongated, shallow profile, which is more efficient in energy dissipation and, hence, flood defence for the hinterland. In this context, dunes are also important. In the same way that upper beaches provide sediment for the lower foreshore in winter, so do dunes supply sand to the beach, with the same effect. Again, this supports the importance of dunes as sediment stores, and not permanently fixed features.

Wave action on beaches is also extensively modified by the angle at which the waves approach. Waves approaching at an angle will wash up the beach, but always return at right angles to the coast (Figure 2.4). As this movement occurs, there is set in motion a series of processes which ultimately enhance the generation of current activity and edge wave formation discussed earlier. There are two types of wave-induced currents: shore-normal and shore-parallel (longshore) currents. These two forms of current, along with edge waves, are responsible for the movement of most coastal sediment, for changes in coastal landforms, and, by implication, for the form of management adopted for any part of the coast. Consequently, although it is not necessary here to understand the complexities of initiation and the physics of movement, it is important to understand the forms of movement which these currents induce.

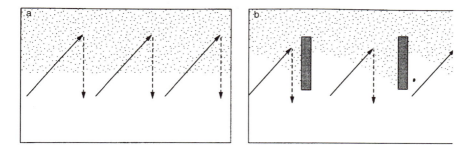

FIGURE 2.4 Wave impact on beaches dominated by shore-parallel currents, and the impacts on the resulting morphology of groynes

Shore-normal currents

Shore-normal currents arise because of the cyclical and elliptical water movements already described. As we have seen, at a certain depth the ellipses fail to complete and a water movement is produced. The ultimate result of this is as the wave breaks, water rushes up the beach (swash), and then retreats again (backwash). The swash and backwash are not necessarily equal in duration, and thus a higher-energy swash will produce a net up-beach sediment movement, while a higher-energy backwash will cause a net off-beach movement. These currents, therefore, move sediment up and down the beach, and onshore and offshore. They vary with the state of the tide, but also seasonally, hence the formation of summer and winter beach profiles.

Further modification also occurs during storms when, because of increased energy, greater volumes of sediment are moved. Because of the higher waves and greater wave base, more sediment falls within the zone of reworking, meaning that sediment can be transported further offshore than normal, with the result that it goes beyond the normal wave base, making it inaccessible to normal wave conditions.

Shore-parallel (longshore) currents

As the name would suggest, whereas shore-normal currents move material up and down the beach, shore-parallel currents move sediment along the beach, producing such phenomena as longshore drift. Such currents are produced by differences in wave height along a beach, or by waves hitting the coast obliquely with the resulting wave gradients mentioned earlier. The resulting current transfers material along the coast, producing significant implications for beach management, as without constant replenishment at the up-drift end, net sediment loss will occur.

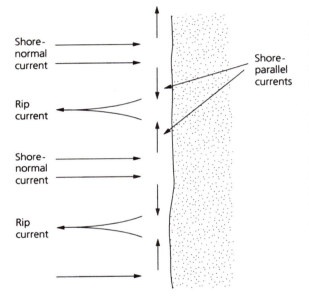

FIGURE 2.5 Interaction of shore-normal and shore-parallel currents to produce rip currents

Combinations

For the sake of completeness, it should also be mentioned that the above forms of current can interact, producing both long-shore and shore-normal movements along the same stretch of coast. This combination, along with edge waves, sets up a series of circulation cells with sediment movement onshore, then along shore, then back out to sea (Figure 2.5). Where the currents move offshore there are formed a series of off-shore currents known as rip currents. Such currents can move a lot of sediment offshore, where it may either be stored or moved back onshore with the rest of the circulatory pattern. It should also be said that rip currents are dangerous to swimmers who get caught up in the rapid offshore movement of water. (A good tip, should this ever happen to you, is always to swim along the coast and not towards the beach, so that you swim out of the current, rather than against it.) The give-away evidence for these nearshore circulation cells is the presence of beach cusps (Plate 2.1).

PLATE 2.1   Beach cusps formed as a result of a combination of shore-normal and shore-parallel currents to form rips (Cayton Bay, Scarborough)

◆   ◆   ◆

It is easy to discuss all the above mechanisms in isolation. In reality, they form a complex series of processes of continuously changing intensity, which are constantly acting on wave-dominated coasts. In addition, so far I have said little about either sediment accretion or sediment erosion. Each of these processes may result in either sediment build-up or sediment removal. Which of these actually occurs is a function of the levels of energy involved, the types of sediment, and the presence of any sea defence measures or other interference by humans.

Figure 2.1 details the processes which we have been discussing. Whenever sediment is transported, the shoreline is changing. If sediment is moved from a beach, that beach must change its form. Thus the shoreline morphology is constantly changing, and as morphology is one of the prime controls on current activity and sediment movement, so the properties of current activity, wave depth, etc., are also constantly changing. As most of the wave energy is dissipated on the lower parts of the beach, it is here that most of the changes occur.

◆   ◆   ◆

So far we have discussed processes as they occur within the nearshore zone, and I have hinted at their implications for sediment. I have said that waves more frequently break along a coast at oblique angles than they do parallel to it. This angle of approach results in the net movement of material along the beach and can result in the movement of material from one end of the beach to the other. In cases where sediment is being constantly replenished at the up–drift end, this movement will be continuous, and so no detrimental effects will be experienced. However, where inputs have ceased, or have been reduced, the sediment deficit will be made up by the removal of material from the beach itself, resulting in a gradual loss of the beach further and further down-drift. In order to prevent this loss of beaches, particularly in areas of tourism and strategic land use in need of protection, groynes are often installed to trap sediment before it moves too far along the beach (Figure 2.4b). Doing this prevents the longshore movement of sediment, but it can mean that places down-drift experience a reduced sediment supply, and as a result begin to suffer from coastal erosion (see Box 3.2, p. 77).

If we assume that there is no restriction on longshore sediment movement, then the continued drift of sediment will result in the formation of new coastal features. Spits and tombolos are two forms of beach which rely on the longshore movement of material for their existence. With the continued movement of sediment along a coast, these features will develop in the mouths of rivers, such as Spurn Head at the mouth of the Humber, or Orford Ness in Suffolk, eastern England; or where the coastline changes direction, such as in the example of Little Sarasota Bay on the Florida Gulf coast. As a result, these features can often provide the type of shelter necessary for much of the sediment deposition and human activity which goes on in estuaries. Problems of erosion can result where the supply of sediment is reduced or cut off following the stabilisation of a beach with groynes, or the defending of a stretch of eroding cliffs (Chapter 3).

Cliff processes and human interference

Cliffs are perhaps one of the most striking types of coastal landform in that they offer visitors the aesthetic attraction of coastal scenery and views. In essence, cliff formation is integral with that of shore platform development mentioned earlier. As the platform develops, so the cliff will remain active. Cliffs, however, occur in different forms, the form being a function of the geology of the area. Geology is important because it governs the rock type and, therefore, the resistance to erosion, and also the geological structure will govern the relationship between rocks of different types.

Taking the Cornish coast in south-west England as an example, tall cliffs composed of metamorphic rocks, known as hornfels, and also granites are both very hard and resistant to erosion. As a result, these cliffs tend to be steep (vertical) and stable. The current wave climate has little effect on them even when wave energy is high during storms, and, as a result, the cliff retreat rate is effectively zero. Because they offer great resistance to wave activity, they provide little material to the coastal sediment budget of the area, and therefore are minor contributors to beach material. From a human perspective, cliff-top developments tend to be safe from collapse and there is little pressure for sea defences at the base of cliffs. As a result, extensive sections of these coasts are 'natural'.

In marked contrast, the cliffs along the Holderness coast or the Suffolk coast, both in eastern England, are composed of poorly consolidated glacial material. Although still vertical, the cliffs here offer little resistance to wave attack and readily crumble into the sea. The cliffs along the Holderness coast are very active and are retreating by between 1.2 and 1.7 m yr^{-1} (Clayton, 1989), but locally the rate reaches 8 m yr^{-1} (Pethick, 1992). They are, in fact, thought to be the most rapidly eroding cliffs in Europe. A similar example can be found at Newport, Oregon, where large-scale erosion since the 1860s has resulted in rapid cliff retreat and loss of people's homes. Here the cliffs are composed of alternating mudstones and siltstones, dipping seawards at around 30 degrees (Sayre and Komar, 1988). As a result of this rapid erosion, their contribution to the coastal sediment budget is great, providing much material for beach build-up. However, despite this important role in the supply of beach material, the very fact that they are eroding at fast rates means that property on top of the cliff is often under threat of falling into the sea, and because of this, public pressure for defence works to be installed can be great (see Chapter 3). It is estimated that around thirty towns and villages have already been lost over the past few centuries along the Holderness coast between Flamborough Head in the north and Spurn Head in the south, while the history of the Suffolk town of Dunwich details the progressive loss of this village (Parker, 1978). In the case of Newport, Oregon, building continued into the 1980s, including development on slumped blocks. By the end of the 1980s further cliff recession had caused much of this new development to fall into the sea (see Sayre and Komar, 1988, for a full account; or Viles and Spencer, 1995, for a summary). Such a degree of coastal erosion occurring at rapid rates in areas of development highlights one of the key problems in coastal management. Clearly, there is a case for defending this coast, considering the threat to property. The dilemma is that the erosion of these cliffs provides such an important supply of sediment that to stop the erosion through defences would mean a serious deficit to the sediment budget, with the likelihood of initiating erosion down-drift and thereby generating new threats to many more properties.

The examples of Holderness and Dunwich cited above are basically coasts with uniform rock type. The vast majority of cliffed coasts, however, resemble the example from Newport, and contain a series of different rock layers, each with its own geotechnical properties. In such situations, the strength and resistance of the coast to erosion is a function of the weakest rock type present, because this is the layer which is going to be most easily eroded, and is most likely to fail. Hence it can be said that a cliffed coast is only as strong as its weakest part. This weakest part can be manifest in a series of ways.

Relationship of structure to the coast

Cliffed coasts can be said to be either concordant or discordant (Figure 2.6). Concordant coastlines, where structure lies parallel with the coast, will form embayments in which wave refraction will play an important control in bay form and erosion

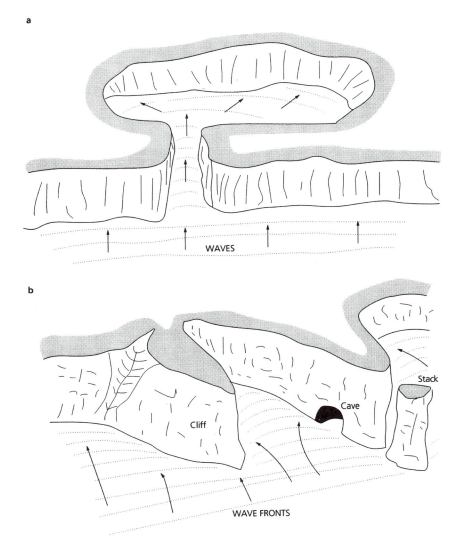

a

WAVES

b

Stack

Cave

Cliff

WAVE FRONTS

FIGURE 2.6 Typical concordant (a) and discordant (b) coastlines

pattern (see Figure 2.3). Discordant coastlines, where structure lies perpendicular to the coast, take the form of a series of headlands and bays, again with wave refraction patterns being important.

Combinations of strata

Many cliffs are not composed of single rock types, but contain more than one. These are referred to as composite cliffs. From a management point of view, they are sometimes difficult to handle. The exact shape and form of the cliff will depend on such factors as strength of the waves, strength of the rock, structure of the rock, and relative hardness of different rock types.

Figure 2.7 demonstrates several cases where the relationship of rock strength to erosive power varies. In Figure 2.7a, resistance to erosion in all cases is much greater than the erosive power, therefore the cliff will retreat uniformly at a rate determined by the lithology; for granites the rate will be slow, for tills fast. In Figure 2.7b, resistance to erosion is much less than erosive power. It is clear that the resulting profile varies accordingly. In addition, different lithologies also serve to modify the profile. Naturally, if the cliff is composed of different lithologies, then different parts of the cliff will retreat at different rates.

All the cases in Figure 2.7 deal with horizontal strata. But what if the strata are not horizontal? Figure 2.8 highlights some other possibilities. Figures 2.8a and 2.8b show dipping rocks. In Figure 2.8a, the direction of dip is inland, and, with the bedding planes representing the greatest plane of weakness, sliding is unlikely as all the forces of movement are directed into the cliff. In contrast, in Figure 2.8b, the direction of dips is seawards, meaning that there is nothing to stop the rocks sliding. Hence, in the cases of dipping strata, seaward-dipping rocks pose greater management threats than do landward-dipping rocks.

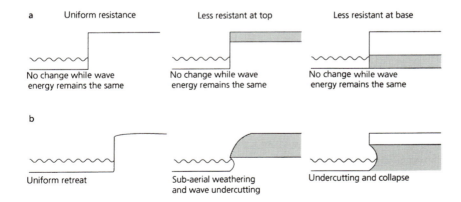

FIGURE 2.7 Effect of different relationships of resistant and less resistant rocks in cliff morphology

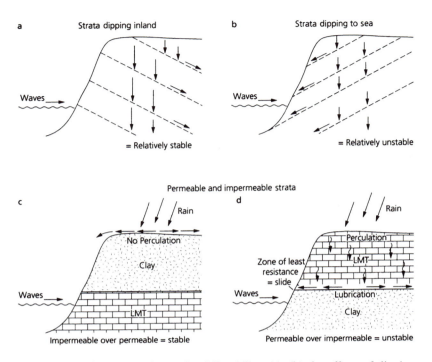

FIGURE 2.8 Controlling factors in cliff stability: (a), (b) the effects of dipping strata; (c), (d) the effects of permeable and impermeable strata

Figures 2.8c and 2.8d represent another scenario. The relationship of permeable and impermeable rocks is also important, as this controls which rocks water can pass through and which it cannot. Where water can pass through one layer and not the one below it, then its only courses of action are to accumulate, building up pore water pressure, or to flow out along the bedding plane between the two layers. In either case, the likelihood of failure is increased. In Figure 2.8c, the fact of impermeable rocks lying over permeable ones means that water cannot penetrate the cliff from above, and so no downward movement occurs. If the impermeable rock is jointed, then any water which does soak in can naturally soak away through the permeable rock beneath. In the opposite extreme, Figure 2.8d demonstrates the movement along the boundary between the two rock types, leading to a situation where failure is likely.

◆   ◆   ◆

Each of the above scenarios indicates natural situations. It is because these situations pose varying threats to the human usage of the coastal zone that we have interfered to try to prevent them happening, and it is because of these threats that we protect cliffs by preventing wave activity from reaching them, so causing the majority of the coastal problems with which this text deals. By defending cliffs, you may achieve your goal of stopping erosion, and thus safeguarding the cliff-top properties, but you also achieve another result, in that you also remove the cliff from the sediment budget, effectively cutting the beach off from its supply of sediment. As a result of this, new erosion can be initiated down the coast, transferring the problem elsewhere (Chapter 3).

Tide-dominated coasts

In more sheltered areas, often where the intertidal and subtidal profiles are sufficiently shallow, wave action is largely removed, either by a process of shoaling or by the direct shelter of a river mouth by a spit. In such cases, waves cease to be the dominant influence, and tides take over as the main formative process. As a result, the depositional landforms reflect the change in dominance, and associated decrease in energy levels, by producing a series of different intertidal habitats. Because of the lower energy levels, sediments tend to be dominated by silts and muds. The deposition of these sediments is governed by the tide height, and the position of the depositional environment relative to the tidal frame. Tidal range – that is, the difference in elevation between high and low water marks – can be divided into three classes:

Microtidal:	< 2 m
Mesotidal:	2–4 m
Macrotidal:	> 4 m

The significance of tidal range lies in the fact that intertidal habitats, such as salt marshes, will occur only in areas which the tide can cover. Also, because the areas at the highest elevations are covered only for a short period around high tide, they receive less sediment than other areas lower down. Hence the higher a salt marsh gets, the less sediment it receives and the slower the rate at which it accretes. Studies done in the Severn estuary (French, 1996b) show that the three distinct salt marsh surfaces which occur here are accreting at rates of 12.1 mm yr^{-1} on the lowest marsh (which is covered more frequently by the tide and, therefore, receives more sediment), 6.4 mm yr^{-1} on the middle-level marsh, and 2.3 mm yr^{-1} on the highest-level marsh (which is covered only by high spring tides and, therefore, receives little sediment) (Figure 2.9).

Most people will be familiar with tides as being those events which cause the sea to rise and fall twice a day, constituting the tidal cycle. This, however, is not the only form of tidal cycle there is. As well as there being fluctuations twice a day, every fortnight there is also a fluctuation in which high tides first increase in height and then decrease, being referred to respectively as spring and neap tides (note that 'spring' has no seasonal inference here). In addition to this, in the spring and autumn, the fortnightly spring tides reach their maximum, the minima occurring in the summer and winter; hence here we do bring in a seasonality. There is also a fourth level of fluctuation which occurs over longer time periods still, such as annual cycles, and the maximum 18.6 year cycle, at which the highest tides of all occur.

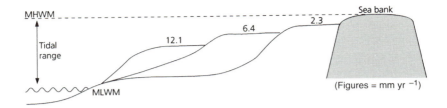

FIGURE 2.9   Effect of frequency of tidal inundation on the vertical accretion rates of recent salt marshes

MHWM = mean high water mark; MLWM = mean low water mark

Hence, tidal heights fluctuate on several scales:

◆   twice a day (high and low tide);
◆   fortnightly (springs and neaps);
◆   seasonally (equinox tides);
◆   every 18.6 years (nodal cycle – the last nodal cycle maximum was in 1988/89).

The physics of these tidal relationships is complex, but is all interrelated and associated with the influence of lunar processes and wave motion. We can view the tide as a long-period wave (see the section on wave-dominated coasts), where high tide equates to the wave crest and low tide to the trough. These waves are subject to external influences, as are gravity waves (see the section on waves).

There are other important aspects of tides which also need consideration. When tides enter shallow water, which in effect is what they do when they enter the nearshore zone, or an estuary, they can also become influenced by short-period harmonics caused by local topography: the tide wave becomes distorted by the inter-action with the longshore profile, local freshwater flows, and estuarine morphology. The effect of this is that the normal flood/ebb tide period of 12 hours can become unequally divided, such that the normal pattern of a 5.5 hour flood, 1 hour high water, and 5.5 hour ebb (symmetrical tide) may become asymmetrical with the flood tide lasting 3.5 hours and the ebb 7.5 hours. This process is referred to as tidal asymmetry, and is important in coastal management for the following reasons:

1   The tide moves a set amount of water into and out of an estuary on each tidal cycle. If the flood and ebb tides are of unequal length, then on the shorter, the fixed volume of water has to move faster than on the longer. That is, if the flood tide takes 3.5 hours and the ebb tide 7.5 hours, then the estuary has, in effect, 3.5 hours to fill up and 7.5 hours to empty. The only way that this can be achieved is by variations in velocity and, therefore, energy. If, as in the situation above, flood periods are shorter, and energy greater, more sediment can be carried in on the flood tide than can be moved on the ebb (assuming adequate sediment supply). Hence, the estuary will experience a net input of sediment. In contrast,

if the situation were reversed, and the flood tide took longer than the ebb, there would be greater velocities on the ebb, greater energy on the ebb, and, therefore, more sediment moved and a net sediment loss to the system. Estuaries which have shorter flood tides, and therefore higher-energy flood tides, are known as flood dominant. Estuaries which have shorter, and so higher-energy, ebb tides are known as ebb dominant.

2   This net movement of sediment is also modified by other factors, such as residual currents, the transport mechanism of sediment transport (bedload or suspension), sediment stability, and sediment entrapment.

Once inside an estuary, tides undergo significant modification as frictional drag is increased owing to channelisation of flow, and also increased bed roughness, processes similar to the attenuation of waves across the foreshore. This can cause a variety of effects, one of the most dramatic being when the estuary cross-section forms a profile not dissimilar to that of a funnel. Here, the channel shallows and narrows at the same time, and so the only way to accommodate the volume of water is to increase its height. On spring tides, this height increase can be sufficiently great to produce a tidal bore (Plate 2.2).

Second, tidal height can also be affected, positively or negatively, by storm activity. A deep low-pressure system (depression) centred offshore at high tide can cause increased water elevation at high tide. The magnitude of this effect is of the order of 0.01 m of increased height for each millibar of decreased/increased pressure.

PLATE 2.2   The combination of high tidal range and rapid narrowing and shallowing of an estuarine channel resulting in the formation of a tidal bore wave (Severn estuary, UK)

These variations are referred to as meteorological tides or surges. A surge tide can be a significant event and cause major disruption and damage to coastal communities. In 1953, a northerly gale generated by a deep depression centred over the northern North Sea caused water levels in the southern North Sea to rise by 3 m above those predicted. This caused widespread flooding up to 6 km inland due to overtopping of sea walls and general sea defence failure. The direct impact on the human population was catastrophic, with the death of 307 people in eastern England and 1,800 people in the Netherlands. In addition, over 100,000 people had to be evacuated from the affected area (Grieve, 1959, in Hansom, 1988). While this illustrates the severity of storm surges, it is certainly not the only example. Hansom (1988) also reports a surge in the Bay of Bengal in 1970, which drowned 700,000 people and modified large parts of the coastline in one night. In the Gulf of Mexico, hurricane-induced surges also cause great damage, while storms affecting the Atlantic seaboard cause severe damage and widespread flooding (Dolan and Davis, 1992).

Earlier, when talking about waves, we saw how wave activity can be significantly modified by current activity, and that such activity can have significant implications for sediment movement. In a similar way, tides also become modified by currents, although tidal currents are distinct from wave currents in that they are controlled by the rising and lowering of the tides. On open coasts currents are at a maximum at high and low water, reversing at mid-tide. In estuaries, currents reach maximum landward velocity at mid-flood tide, decrease to zero at high and low tide, before reversing to maximum ebb velocity midway through the ebb. This tidal current relationship is of fundamental importance in determining intertidal morphology, because where the currents fall to zero the finest sediment can settle out and become deposited (see Figure 2.10).

Although we clearly have a range of tidal cycles, operative for different time periods, the main input and output of sediment, and corresponding morphology, is controlled by diurnal and semi-diurnal tides. These tides may occur once a day (i.e. one high and one low water during each 24 hours – diurnal), as happens along much of the Antarctic coast and parts of Australia and New Zealand, or occur twice a day (semi-diurnal) as along the Arctic and Atlantic coastlines; or there may be combinations of these two types, such as on the coasts bordering much of the Pacific Ocean. In some areas, however, such as the Solent in southern England, high and low water can occur four times a day.

Tidal flow velocity varies with the state of the tide, and consequently the material deposited during different parts of the tidal cycle will vary. Where the tidal currents are slowest, fine clays and silts can settle out to form mudflats. Where currents are higher, these fine particles may be kept in suspension, resulting in more silty and sandy deposits (Figure 2.10). This depositional pattern results in the characteristic sediment distribution often seen in estuaries and sheltered embayments, of sandflats, mudflats, and salt marshes. The exact form of deposition is extremely varied, but a summary of key types is presented in Figure 2.11.

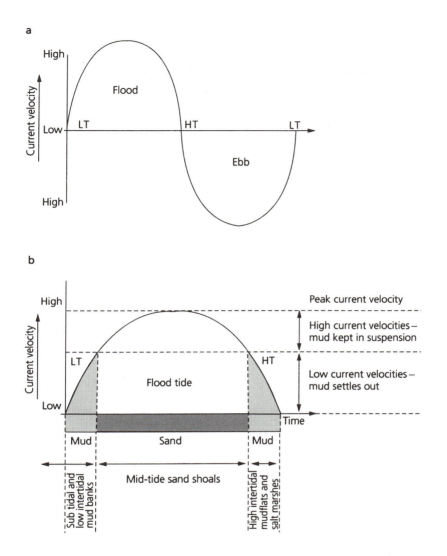

FIGURE 2.10 Relationship of current velocity to intertidal depositional environments: (a) diagrammatic illustration of the variation in current velocity over a complete tidal cycle; (b) sediment distribution in the intertidal zone in relation to flood tide current velocities

HT = high tide; LT = low tide

Source: French, 1990

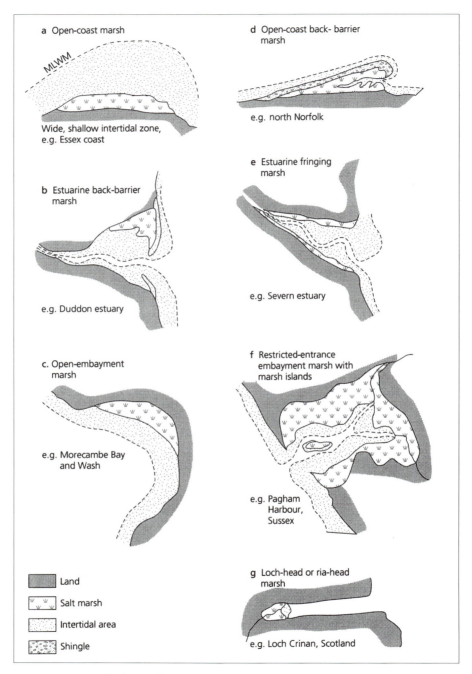

a Open-coast marsh

MLWM

Wide, shallow intertidal zone,
e.g. Essex coast

b Estuarine back-barrier
marsh

e.g. Duddon estuary

c. Open-embayment
marsh

e.g. Morecambe Bay
and Wash

d Open-coast back- barrier
marsh

e.g. north Norfolk

e Estuarine fringing
marsh

e.g. Severn estuary

f Restricted-entrance
embayment marsh with
marsh islands

e.g. Pagham
Harbour,
Sussex

g Loch-head or ria-head
marsh

e.g. Loch Crinan, Scotland

Land

Salt marsh

Intertidal area

Shingle

FIGURE 2.11 Variation of salt marsh types in a range of tidally dominant environments
Source: Modified from Pye and French, 1993

Mudflats and sandflats and human interference

It is important to bear in mind that sediment will be deposited only once it has become too heavy to be carried in the water. Hence sand will settle out from water which is faster-flowing than that needed for mud to settle out. Each tidal cycle has two components, a flood tide (tide coming in) and an ebb tide (tide going out). The velocity at which the water flows varies during this cycle according to the graph in Figure 2.10. The periods of low velocity occur when the tide is turning, and, theoretically, the flow should actually be zero as the direction of flow changes. What this actually means is that tide-dominated environments often contain a sediment sequence from the shoreline to low water channel which goes from mud to silts/sands to muds. In situations where the low water channel migrates away from one shore-line, or where an estuary is being infilled with sediment, this sequence may also occur vertically, although estuarine stratigraphy is a potentially complex topic.

It can be seen that where tides dominate, sediments settle out in discrete areas. However, if we think about the settling of mud, typically material composed of fine clay material, then under laboratory conditions this material could take many hours, or even days, to settle out in pure water which is completely undisturbed. Under experimental conditions, the rate at which settling occurs is given by Stokes's law, which states that the settling velocity of a particle is related to its diameter and the temperature of water through which it flows. It is clear from this law that settling of fine clay particles cannot occur during the few hours around low and high tides, when the water is disturbed. Contrary to this, however, we do find extensive areas of fine clay in mudflats, so clearly there has to be some other process in operation which allows these finest particles to settle out.

In solving this problem, we come to the second important role of the tides. Estuaries and shallow embayments are often associated with river mouths; that is, areas where fresh water enters the sea. When the tide comes in, this fresh water mixes with the salt water brought in from the sea. In its simplest sense, the mixing of the two types of water causes electrolytic imbalances to occur, which means that clay particles become attracted to each other and adhere together. As more and more particles stick together, the effective size and weight of the particle cluster increases and so is able to settle out as if it were a larger silt or sand grain. This process of particles sticking together is known as flocculation, and without it we could not have the large areas of mudflats which typify many of the world's estuaries. Because of the large quantities of mud which settle out, an area can contain large expanses of intertidal flats, which become very important as habitats and of conservation importance. In addition, as with beaches, a wide, shallow intertidal mudflat is very efficient at dissipating wave energy, and is therefore an important form of coastal defence in its own right.

In addition to the conservation and coastal defence potential of estuaries, from a human perspective the sheltered nature of these environments makes them particularly appealing for development, albeit for leisure activities, such as marinas, or for commercial port activities (see Chapters 3 and 4). The wide expanses of inter-tidal muds also provide ideal areas where land can be claimed for new quays, etc., reaching down to the low water mark. As a result, many such areas have come under great threat, both historically and up to the present, as an ideal environment in which to claim and develop land. The outcome of this is that most estuaries have experienced large-scale claiming of intertidal land which has completely modified the shape of the estuary, with further implications for channel process, flood/ebb tide dominance (see Chapter 3), and also for wildlife which use such areas for feeding (see Chapter 4).

Salt marshes and mangroves and human interference

As mud accumulates according to the processes discussed above, so the mud surface increases in elevation, and as it does so, it becomes covered in water for shorter periods of time owing to its changing position relative to the tidal frame. As these new conditions develop, so some pioneer plant species start to become established, such as *Spartina* sp. or *Salicornia* sp. From this, we start to get the development of saltmarshes. In essence, salt marshes are just mudflats with vegetation on them. Different types of plant can tolerate different amounts of salt water, and so the lower parts of the mudflat will have a flora different from that of the upper parts, and as the marsh develops, so newer species will start to colonise, until ultimately there will be a transition between the salt-tolerant marsh and terrestrial vegetation. However, such transitions are rare because human intervention normally means that some form of defence or reclamation work has occurred before this stage is reached (Chapters 3 and 4).

Once vegetation is established on mudflats, the whole process of sedimentation changes because the vegetation acts, first, as a baffle, trapping sediment in its leaves and among its roots, thus increasing the rate of sediment build-up, and, second, as a means by which the sediment can be bound together, with the roots helping to hold the sediment in place. By such processes do salt marshes develop, but these natural processes are subject to interference by such activities as dredging (increasing suspended sediment and potentially producing a swamping of marsh vegetation), tidal reservoir construction (prolonged submergence of marsh vegetation beyond that normally tolerated by species and reduced tidal range), sea defence (preventing landward transitions and also preventing the natural landwards movement of species during sea level rise, leading to coastal squeeze), and pollution (potentially toxic effects of various pollutants, and also the storage of pollutants in salt marsh sediments) (Chapter 4).

Marshes take on a whole set of characteristics once they have begun to form. Creeks develop, vegetation spreads and surface features such as salt pans also form. Another feature of marshes is that they are very productive biologically, and so they have often been seen as ideal sites for conversion to agricultural land. As a result, huge areas of marshes have been claimed and drained to produce fertile agricultural land. This has two direct implications. First, there needs to be a form of coastal defence to prevent the sea from reflooding the area, and second, there is a dramatic loss of intertidal area, resulting in constricted estuarine channels, artificially steepened profiles, and modifications to the tidal prism (Chapter 3).

Whereas salt marshes are distributed throughout the temperate climates of the world, in tropical areas the intertidal mudflats become colonised by mangrove trees (salt-tolerant evergreens) whose roots and stems provide the same function as do the smaller versions on salt marsh plants. In development and process, they are similar environments, although, owing to their higher levels of organic material, soils in mangroves tend to be more peaty than muddy. In addition, the ecology of the two environments is such that their usage varies. Salt marshes are dominated by low-level flora, suitable for grazing or turf cutting. Mangroves, on the other hand, are dominated by trees, and thus are more suited to logging.

Mangroves have declined significantly in area, especially in Asia and Africa, with the overriding causes being lack of protection and management, largely driven by the increasing population, causing large-scale conversion to agriculture, aquaculture, urban/industrial developments and fuel wood. In Guinea, Guinea Bissau, and Sierra Leone, conversion of mangroves to rice fields and the extraction of fuel wood have accounted for 46 per cent of the original mangrove area (Middleton, 1995). In the Philippines, 75 per cent of the mangrove cover has been lost over the past seventy years, originally as a result of fuel wood collection, but subsequently through the creation of fish ponds for aquaculture. The destruction of mangroves and their conversion to other uses are not really justified on economic grounds. Christensen (1983) demonstrates that the income from mangroves converted to rice cultivation is of the order of US$168 per hectare per year, while an undamaged hectare used for shrimp farming can yield of the order of US$206 per hectare per year. This costing also ignores the cost of the natural coastal protection afforded by the mangroves, thus making it even more cost-effective to keep mangrove protection.

The largest remaining areas of mangroves occur in the Ganges and Brahmaputra deltas, where they have been protected by reserve status since 1875.

Coral reefs

Coral reefs tend to occur in warm, tropical, shallow waters, typically between 30° north and south of the equator. Their importance is exemplified in the fact that although they occupy only 0.2 per cent of the ocean floor, they are incredibly diverse habitats, containing around 25 per cent of the world's marine species (Middleton, 1995). Apart from the habitat importance, coral reefs form natural offshore break-waters (see Chapter 3) and so play an important role in the protection of the coast from waves and, as a result, from coastal erosion.

Reefs are particularly susceptible to environmental impacts, both natural, such as hurricanes, and anthropogenic. Many of the problems relating to reefs are due to the impacts of tourism (see Chapter 5), but others occur indirectly, because of the increased amounts of sediments derived from rivers (Chapter 6), coastal development, pollution, and also military activity, such as nuclear weapon testing (Bikini Atoll) and conventional military ranges (Kwajalein Atoll, Marshall Islands) (Middleton, 1995). Because of the distinct link with tourism, many of the issues which stem from this source are discussed in detail in Chapter 5.

Wind-dominated (aeolian) coasts

Although not strictly separable from wave- and tide-dominated environments, wind-dominated (or aeolian) coasts do not rely directly on the waves or tides in order to form (they do, however, rely on these agencies for the delivering of their sediment in the first place). The difference between dunes and sandy beaches is that in the former, wind has transported the sand from the intertidal zone into the supratidal zone. Hence although important in a coastal context, aeolian coasts are not dependent on inundations of salt water for their formation or the survival of their plant species. The primary coastal landform associated with aeolian coasts is sand dunes. In order to form, dune coasts need both a supply of sand delivered to the beach and a strong onshore wind.

In order for sand to be transferred into the dune system, the wind has to achieve a suitable velocity to initiate motion in the individual grains. This initiation is a function of two things. The first is the velocity of the wind. Wind speeds of around $4\,\mathrm{m\ s^{-1}}$ can induce some movement in dry sand, although speeds of around $20\,\mathrm{m\ s^{-1}}$ and higher are needed to achieve rapid sand movement (Bagnold, 1954). Second, the water content of the sandy beach is important. The wetter a beach is, the more cohesive strength it has, and the harder it is to initiate movement in the grains. In effect, this means that tidal ranges are important as the sediment needs to be exposed to drying for long periods of time. For example, dune coasts occur where there is a large tidal range and a foreshore which is wide and shallow. This exposes large areas of sand for long periods of time, and so allows the sand to dry and be blown into the dunes (Plate 2.3).

PLATE 2.3   A wide intertidal sandflat fronting a dune complex (North Norfolk, UK)

Deposition of sand will occur only when either the wind velocity drops, or the grains are prevented from further movement by hitting an obstacle. As soon as either of these occurs, accumulation will start.

In order for the system to become stabilised, there is also a necessity for vegetation to trap and retain sediment. Typically, this needs to be drought tolerant, a good example being marram grass. Dunes are highly fragile environments, although they play vital roles in the stability of the beaches fronting them. Their fragility is due to the uncompacted sediment and poor binding of the sediment by roots. Damage to dunes is, therefore, easy to cause and largely arises from the impact of humans walking over them. Furthermore, their aesthetic appeal means that they are often the site of intense visitor pressure. Being in essence piles of sand, they are very easily eroded and disturbed (see Chapter 5).

Deltaic coastlines

Whereas the coastal landforms discussed above can be attributed to one dominant formative process, deltas can be both tide- and wave-dominated, as well as river-dominated, and can equally fit into either of the above groupings. Their origin lies in the fact that some rivers discharge sediment into the sea which is not completely reworked and so builds up in the river mouth and protrudes out from the coastline into the sea, as a sedimentary deposit. The distinction of dominant process lies in their shape, and deltas tend to be labelled according to a type locality; that is, according to the place where the typical formation is shown (Wright, 1978).

Where tide and wave conditions are low, the river discharge dominates and the delta is characterised by many channels. This type of delta is traditionally referred to as a birdsfoot or Mississippi-type delta The sediment accumulates where it is deposited and little is reworked by waves and tides. Where river processes are subordinate to either waves or tides, the sediment discharged by rivers is reworked and refashioned by coastal processes, hence the form of the delta owes more to these secondary processes than to the river.

Where the coast is dominated by tides, with little wave activity, the delta is typified by shoals (small islands) surrounded by water or swamp vegetation, often with offshore barriers, or bars. These are typically known as lobate or Nile- or Niger-type deltas. Where deltas form in conditions of high wave energy, much of the sediment is reworked and redistributed around the river mouth by a process of smoothing and rounding. The deltas typically have high, broad, sandy beaches, and deflected river mouths. These types are known as blunt and straight (or alternatively São Francisco- and Senegal-type) deltas.

Deltas are constantly changing, owing to their constant sediment input, and as a result, new material is constantly being added to the seaward edge and to the delta surface by flood waters. Because they owe much of their material to rivers, the fertile sediments are attractive to farmers, leading to large areas of deltas being reclaimed. Many are also of sufficient size to allow settlement to proceed, although, being low-lying, these settlement areas are often threatened by factors such as rising sea level.

In addition, delta channels are highly mobile in that they regularly switch, abandoning part of the delta. Once this occurs, the abandoned area receives less sediment from channel overflow, and so can start to erode. Should this area coincide with development, pressures to defend are great.

Summary of Chapter 2

We can see that the form of any particular coastline is not just one of chance. Each type of coast is the result of a series of processes, and typifies a particular environment. Each type of deposit or landform which occurs is controlled by a set of environmental criteria. These criteria, however, are themselves subject to constant change, and so are themselves also changing. Take for example wave-dominated currents. Wave currents are partly controlled by the morphology of the intertidal profile, such as wave base. However, currents interact with the sea bed, moving sediment around and depositing it elsewhere, and so we have a situation where intertidal morphology controls the process, but the process is constantly changing the intertidal morphology and, as a result, the process itself has to change.

Perhaps the most important message to take away from this chapter is that the coastline is composed of a series of morphologies, whether dunes, marshes, shingle, or sand. In the long term, depositional and erosional patterns can change, so none of these coastal features should ever be considered as being permanent. Increasingly, dunes, marshes, or whatever, must be viewed not as long-term coastal amenities, but as temporary stores of sediment. These sediment stores have been left there by the sea because at the present time the sea does not need them. In the future, there is every likelihood that the sea will need them again, and when that time comes, the sea will begin to erode them. This is a perfectly natural process. We start meeting problems when, because of development and land use, it is just not possible to allow the sea to do what it wants. However, like an incredibly spoilt and manipulative child, the sea cannot be tempered, but will get its own way in the end; the means of achieving this may change, but the outcome will be the same.

The intention of Chapter 2 has been to give the reader a background into the ways in which waves and tides behave, how this is reflected in the resulting landforms, and how it is possible to interfere with these processes to initiate long-term coastal problems. The reader should leave this section with an understanding:

◆ of how waves and tides interact with the environment to produce a series of coastal morphologies;
◆ that coastal features are transient, and should never be considered as permanent;
◆ of the fragility of the coastal process/landform system, and how a relatively simple act can interfere with a process to cause a series of knock-on implications for the rest of the coastline.

It has not been the intention of this chapter to give a detailed account of coastal geomorphology. Such aspects are covered well in other texts. It has intended to highlight those areas where interference with processes can produce human impacts on the coast. It is these points which are now developed further in Chapter 3 (coastal defence and land claim), Chapter 4 (industrial development), Chapter 5 (tourism), and Chapter 6 (indirect coastal impacts).

Chapter 3

Land claim and coastal defence

Introduction

In many papers and texts, authors refer to the process of 'reclamation' when discussing land claimed from the sea. Increasingly, however, the more technically correct term 'land claim' is becoming preferred, thus removing the misconception that reclaimed land represents areas which were once dry land but had been inundated by the sea, and are then subsequently 're-claimed'. While this may well be true in some cases, it is certainly not always the case.

The topics of land claim and coastal defence are discussed together because, in general, they tend to occur together; that is, when undertaking land claim, it is essential to build sea defences. Land claim occurs for two reasons: for the generation of agricultural land, and for development. Development may be related to housing, leisure, or industry, although some aspects occur only in certain areas. For example, land claim for port development is largely, although not exclusively, restricted to estuaries. Ports need to be sheltered, and the open coast is less well suited unless sheltered by a shallow intertidal profile, such as in the case of the port of Heysham, in Morecambe Bay, Lancashire, in north-west England. Apart from port development, land claim occurs primarily for agriculture and industrial development, and this also tends to favour estuaries rather than open coasts, for reasons which are varied but relate to the sheltered nature of the estuarine environment, the supply of large areas of cheap, flat land, accessibility from both land and sea, availability of water supply, and ease of waste disposal.

The term 'coastal defence' is a much used, but much abused term for referring to any feature along the coast or estuary designed to protect a beach or land. It is more correct, however, to distinguish between two forms of coastal defence. First, flood or sea defences are structures used to prevent the land from being flooded by the sea; examples include earth embankments in estuaries. Second, there are coastal protection structures, a term used to describe measures taken to prevent the land from being eroded, such as sea walls along the base of cliffs.

In estuaries, tidal processes facilitate the accumulation of thick deposits of sand, silt, and clay, reaching to the upper limit of the tidal range. The process of land claim means that it is necessary to exclude the sea from part or all of this intertidal area, and also protect this area from reinundation. Land claim for agriculture and industry generally takes in the higher salt marsh because the higher elevation of the intertidal area claimed means, first, that the wave activity will be reduced by the lower marshes and mudflats fronting the area to be reclaimed. In cases where there are not sufficient areas of such deposits fronting the proposed claim, then the area is perhaps not well suited. Second, less material is needed to build up the newly created dry land. Third, the higher the elevation, the lower the sea walls need to be to prevent tidal overtopping. Finally, agriculture needs good-quality farmland, and the upper marshes provide the most 'mature' sediments available in respect of the processes of soil formation.

There are few areas of coastline in the 'developed' world which have not been subject to some form of land claim, defence works, or development. It has been the tendency for coastal populations to utilise their immediate environment to the full, obtaining as much land as possible in order to increase their agricultural or industrial potential, and to defend low-lying land to increase their security from flooding by the sea. Industrial development along coasts and estuaries will be dealt with in the next chapter, while here we shall concentrate on loss of intertidal land and its conversion to dry or freshwater areas.

As far as port development in estuaries is concerned, one of the key requirements is that the port itself should be afforded as much shelter as possible. Traditionally, this meant that estuaries were the most suitable sites and, because of the small size of the earliest vessels, many of the early ports were built some way upstream from the estuary mouth. As ship size has increased, so these earlier ports have become uneconomic, and thus many have been relocated downstream, towards the estuary mouth. This has meant that the majority of the world's major estuaries have some form of port development located within them. In addition, even these areas may not be suitable for the larger vessels of today, and thus many estuaries have been artificially deepened by dredging to allow for the increased draught of modern ships. This process leads to an artificial intertidal profile in which natural processes are in constant competition with the alien environment in which they find themselves. As a result, especially over the past few centuries, many estuaries have experienced considerable modification to their natural ecosystems, which have brought about changes in floral patterns and bird populations. From a process point of view, the dredging and land claiming which have occurred in estuaries have also produced changes in circulation patterns, tidal regime, and sediment deposition patterns, causing further knock-on effects for natural habitats and wildlife populations.

History of land claim

The winning of land from the sea has been an activity which has gone on for many centuries. There is evidence of Romano-British land claim in the Wash and the Severn estuary in the UK, while the Dutch, perhaps regarded as the world leaders in land claim, have extensive tracts of coastal land, formerly intertidal, protected by sea walls which date back several centuries (Goudie, 1981). As well as land claimed for farming, as soon as there was the development of sea-going vessels, there was a need for ports and places to load and unload cargo, which necessitated areas of hard standing and permanent access from the sea. Evidence for such port development can also be dated back to the Romans: Porteus, the great port serving Rome, dates from AD 46 (Singer *et al.*, 1956).

In its earliest times, the process of land claim was carried out largely for the provision of land for agriculture. This was especially true in areas where the hinterland was hilly or rock-dominated and, therefore, unsuitable for cultivation, or where the tribes inland were unfriendly, causing peoples at the coast to want a more secure source of food.

The importance of agriculture was the ultimate driving force behind land claim up to the nineteenth century. It is still of importance, for the following reason. The Industrial Revolution in the nineteenth century which occurred in the UK, and subsequently in other countries, led to a rapid increase in the industrial base of the developed world, and the need for links to other parts of the country and to other countries. As a result, from this time onwards, land claim for port activity started to increase, although the provision of food also increased in importance because more people were migrating away from the rural areas of subsistence farming, to the industrial centres, and so needed to be supplied with food by others. In its earliest stages, industry developed at resource centres (such as coal and ore fields) rather than import/export centres, and so land claim for industry and port development did not really become significant until the late nineteenth and twentieth centuries, when the development of larger industrial plants, particularly petrochemical works, led to a requirement for large areas of flat, cheap land with links to ports. Estuaries were seen as the ideal place for such development and, as a result, large areas of the major estuaries underwent significant land claim, in the UK most notably in the Tees estuary (Figure 3.1), which has lost 83 per cent (c. 3,300 ha) of its intertidal area since 1720, with most of this loss being post-1800 (Davidson et al., 1991; Pye and French, 1993b). The small area of intertidal land which is left (Seal Sands, North Gare Sands, and Bran Sands) is also under threat from developers.

By looking at any major estuary in any developed country around the world, it is possible to see evidence of land claim for industry. Industry, as a result, can be regarded as one of the major factors influencing the shape of estuaries, and also one which has effects beyond physical change, including as it does impacts from pollution, an issue which will be discussed further in Chapter 4.

Although actual figures for land claim are difficult to determine, Table 3.1 details some British land claims which have been quantified. Many of the figures quoted relate primarily to salt marsh and intertidal flats, although other habitats may also be involved in some cases. As well as those estuaries cited in Table 3.1, there are many others, including practically all the main UK systems, which have undergone land claim, but sources do not quantify areas involved. Pye and French (1993b) have attempted some assessment of the extent of reclamation in all UK estuaries, albeit on a qualitative rather than quantitative basis.

Environmental implications of land claim

As soon as any land is claimed from the sea, there is set in motion a series of changes which considerably affect the natural environment. The extent of such changes will vary according to the size of the land claim, and the geomorphic area in which it is undertaken. Claiming a thin strip of upper salt marsh for agriculture would not have the same effect as claiming a block of marsh and mudflats for port development, even

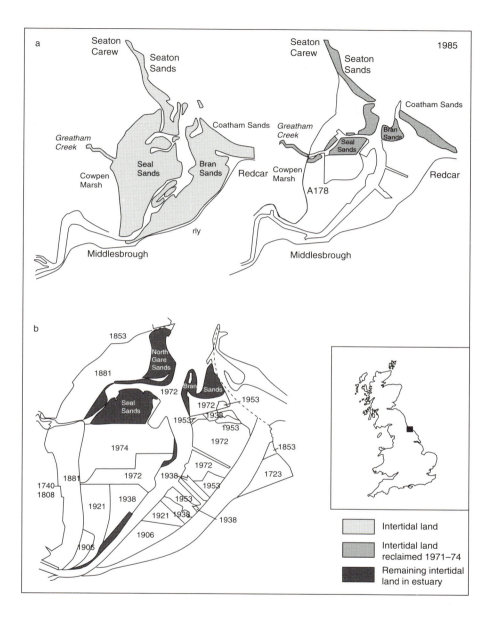

FIGURE 3.1   Impact of land claim on the intertidal area of the Tees, north-east England:
(a) intertidal land in 1861 and 1985; (b) the year of development
Source: JNCC and English Nature, *Colonial Waterbirds* 9

TABLE 3.1   The extent of land claim in some UK estuaries

Estuary	Total area (ha)	Source
Axe	80	Parkinson (1985)
Taw and Torridge	> 480	Pye and French (1993b)
Mersey	> 490 since 1800	Doody (in prep.)
Southampton Water	690 since mid-1800s	Coughlin (1979)
Exe	830	Parkinson (1980)
Orwell, Suffolk	983 since 1200	Beardall *et al.* (1991)
Poole Harbour	> 1,052	May (1969)
Alde, Suffolk	1,450 since 1200	Beardall *et al.* (1991)
North Norfolk coast	1,500	Pye (1991)
Stour, Essex	1,601 since 1200	Beardall *et al.* (1991)
Portsmouth Harbour	1,819 since 1500	Davidson *et al.* (1991)
Deben, Suffolk	2,242 since 1200	Beardall *et al.* (1991)
Ribble	2,320 since 1800	Doody (in prep.)
Tees	3,300 since 1720	Davidson *et al.* (1991)
Dee	6,000 since 1730	Davidson *et al.* (1991)
The Wash	47,956 since Saxon times	Buchner (1979), Hobbs and Shennan (1980), Lincolnshire County Council (1980)
Humber	75,110 since Roman times	Steers (1946)
Severn	84,000 since Roman times	Allen (1991)

Source: Data from Pye and French, 1993b

if the total intertidal area claimed was the same. The reasons for this are varied, but, perhaps most importantly, land claim has significant secondary effects on system hydrodynamics, such as tidal flow, current velocities, ebb/flood symmetry, etc. The very fact that land claim involves the taking of intertidal land from the marine environment means that:

1   The tide will have less area to inundate and so water depth could increase, covering areas for longer, and also increasing the tidal range upstream (Figure 3.2).

2   To prevent the sea inundating the claimed land, sea defences have to be built, which themselves represent a 'hard' barrier which can act as a reflector for waves, causing back-scour, and preventing any natural habitat adjustment in response to external factors, such as sea level rise and dredging. Generally, when land is first claimed from the sea, hard defences are employed; soft defences (habitat creation), in contrast, tend to be used as a means of protecting areas previously defended, or to protect coasts where the sea is threatening to erode existing hard defences.

3   The cross-sectional area and morphology of an estuary will be modified by reducing the areas of intertidal habitats, thus causing further implications for current activity, wave propagation, and sediment movement.

4   It has already been indicated that coastal deposits, such as dunes or marshes, should be regarded not as features fixed in time, but purely as temporary stores of sediment. By enclosing estuarine and coastal deposits behind defences, we are, in fact, removing part of this temporary sediment store from possible reworking, when environmental conditions change, and, as a result, large volumes of sediment may be removed from the sediment store, which may lead to a future deficit in the sediment budget of a system. While the system is in an accretional phase, or one of net stability, the potential problems may not be realised. During erosion, however, problems of sediment starvation elsewhere in the system may well result, leading to new threats to sea defences, or loss of habitat.

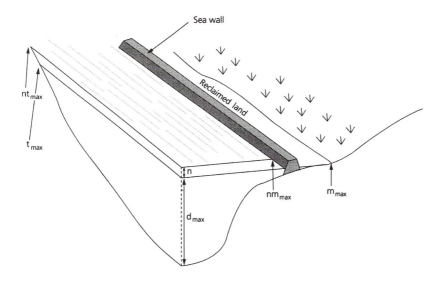

FIGURE 3.2 Impact of land claim and sea wall construction on the intertidal area of an estuary

d_{max} = maximum water depth before land claim; t_{max} = maximum upstream limit of tidal range; m_{max} = maximum limit of tidal range on estuary margin; $n + d_{max}$ = new maximum water depth; nt_{max} = new maximum upstream limit of tidal range; nm_{max} = new maximum limit of tidal range on estuary margin

Despite these environmental problems, land claim has been practised for many centuries because the gains from the newly acquired land have been considered greatly to outweigh the drawbacks of land claim. To take another stance, however, it may be more correct to say that during the periods of greatest land claim, largely due to agricultural expansion, these potential drawbacks were not foreseen and, therefore, not considered. It is only in recent decades, when many of the older defences have started to become undermined, that coastal engineers have become aware of the implications of what our forebears have done, and the legacy of problems which the current generation has been left to solve.

The need for coastal defences

Because of the need to protect the newly claimed land from reinundation by the sea, it is also a requirement of land claim to install some form of flood defence. This defence is largely to keep the sea out, rather than an anti-erosion structure, and so tends to be more a flood defence measure than a measure for coastal protection. Typically this defence is an earth embankment, sometimes faced with stone, although occasionally defences may be concrete walls, particularly in cases of high hinterland value, such as land claim for industrial development, nuclear power stations, or port activity.

These flood defences are often constructed using material derived from the upper salt marshes. Methods of obtaining this material can vary, but in the Severn estuary in the UK, for example, the upper marsh is scraped, and the material piled up to form the embankment. The result of this is an upper marsh largely devoid of vegetation, and lower by a few centimetres. The implications for the system are that vegetation needs to recolonise the area and an equilibrium state be regained before any significant vertical marsh accretion can occur. The process of vertical marsh accretion is essential if a marsh is to keep pace with sea level rise.

An alternative method can be observed on the east coast of England. In the Wash, individual pits (borrow pits) are dug to a depth of a few metres adjacent to where the new defence is to be built. Again, material excavated is piled up to form the flood defences. As compared to the earlier method, digging discrete pits means that most of the upper marsh remains undisturbed, although parts of this are lowered considerably. The methodology states that infilling with water during flood tides will input sediment, and vegetation will begin to colonise, with the end result that the pit will, after a period of around fifteen years, merge in to the rest of the marsh. While this is generally the case, recent work has indicated (Pye and French, 1993b; Osborn and French, in preparation) that in a significant number of pits there is only limited infilling, and, far from being accretional, the pits are behaving as tidal reservoirs (Plate 3.1), discharging large volumes of water on the ebb tide, and causing erosion of the marsh creeks, and internal dissection of the marsh itself.

Clearly, the onset of erosion due to borrow pit construction indicates a failure of the borrow pit methodology. It is possible that the problems in these cases lie in the fact that the pits were dug too deep, and so lie below the level where marsh vegetation can normally colonise. Because of this, any sediment that does accumulate in the pits is rapidly removed by scour, owing to the high velocities reached during the ebb tide. It is also possible that the explanation is more complex than this, including variations in wind/wave climate or the decline in sediment availability due

PLATE 3.1   Large borrow pit fronting a sea bank (Gedney Drove End, Wash, UK)

to coastal defences. Whichever of these explanations, or combination of reasons, is correct, poor construction methods and/or poor understanding of the system hydraulics have produced a situation which is the reverse of that naturally operating in the local environment; that is, localised areas of erosion in an environment of general accretion.

Once the defences are built, where the land is to be farmed it needs to be left to allow the salt content to be reduced sufficiently to allow crops to grow. After being claimed, the new land, previously covered by water twice a day, will start to dry out and dewater, which will itself cause the land to compact and shrink, thus causing the land surface level to fall. After a period of years, the level will fall sufficiently to cause the new land surface to lie below the marsh level to seaward of the defence (which has continued to accrete vertically since the defences were installed), and greatly below sea level. For this reason the flood defences become even more important, especially where large areas of land have been claimed and could potentially be flooded were the defences to fail. As an example, work done for the UK National Rivers Authority (now part of the UK Environment Agency) has indicated that were the defences in the Wash to fail, the area of land flooded (with the exception of a few isolated areas of high land) could potentially reach as far inland as Cambridge, the first extensive land area above sea level. Elsewhere, surveying across sea defences in the Severn estuary (Allen, 1991) has indicated that this variation in height either side of the sea defences can show a relative drop on the landward side by as much as 1.7 m over a period of c. 1,700 years, or 0.88 m over the past 650 years. Such variation clearly demonstrates the ongoing commitment to defence once land has been claimed. The relative height difference is a direct result of a sinking land surface due to dewatering and compaction inland of the defence, and an increase in marsh surface level to seaward due to continued vertical sediment accretion. Visually, this difference can be dramatic, especially during a high spring tide. Such an example can be seen later, in Plate 3.2.

The need for flood defences to protect land claim is also important in an environment of continuing sea level rise. Although rise in sea level occurs eustatically, the effect on the land is very much a regional problem, owing to the fact that coupled with variation in sea level is movement of the land relative to the sea. Although strictly speaking this effect is a natural phenomenon, and so not a result of human influence, the very process brings about many and varied responses from humans in an attempt to protect the developed coastal strip, and hence the process of sea level rise and land subsidence, collectively referred to in terms of relative movement of the sea, and termed *relative sea level rise*, has a profound impact on how we manage the coast and how we interact with it. For example, if we take a natural coast, with no form of defence and a gradual landward transition between salt-tolerant and fresh-water vegetation, the floral change landwards would reflect a trend from salt marsh to freshwater species (for the sake of this example). In conditions of sea level rise, such vegetation zones can 'move' landwards, thus ensuring that they constantly occupy the same position relative to the tidal frame; that is, a particular vegetation zone will be covered by sea water for the same length of time and by the same frequency of tides. The system is thus dynamic and can respond to fluctuations in environmental conditions, while maintaining an equilibrium. This is one of the main reasons for maintaining a natural coastline wherever possible, because it allows habitats to respond to changing environmental conditions. It does, however, result in the loss of land.

All the time the system is in its natural state, any long- to medium-term fluctuation in sea level can be compensated for by landward shifts in vegetation community zones. As soon as sea defences are built, however, any landward shift in these zones is prevented by a physical barrier, and then vegetation communities can be lost by a process known as *coastal squeeze* (Figure 3.3). Without the freedom to migrate landwards, species are fixed with regard to the space which they occupy. As conditions change, particular species become intolerant of the new environmental conditions, such as the frequency and duration of inundation by the tide. In this situation, marsh communities can revert to lower marsh species, meaning those which can tolerate greater and more frequent periods of inundation by the sea. As a consequence of this, with relative sea level rise continuing, the depth, frequency, and period of inundation will increase the possibility that a marsh will actually revert to lower marsh species, or even mudflat. In such cases, the increased depth of water will eventually allow greater wave activity and increased threat of overtopping of the defences and subsequent flooding.

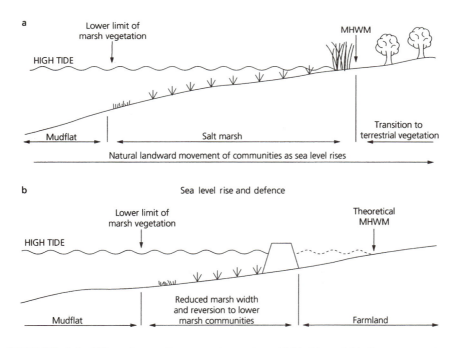

FIGURE 3.3 Effect of sea wall construction on intertidal habitats: (a) before construction; (b) after construction

MHWM = mean high water mark

Modification of the tidal prism

The disruption of the tidal regime of an area by human interference, especially land claim, can produce major modifications to the hydrodynamics and geomorphology of the coastal or estuarine system. The distinction of tidal range has been introduced in Chapter 2, with the greater tidal ranges being responsible for the movement of greater volumes of water. The volume of water exchanged on each tidal cycle – that is, the difference in water volume between low tide and high tide – is referred to as the *tidal prism*. Carter (1988) suggests that a crude way of estimating this volume is by measuring the surface area of the water at high tide, and multiplying by the tidal range.

The volume of water exchanged during a tidal cycle occurs through flood and ebb tides, leading to estuaries being classified as either flood dominant, ebb dominant, or co-dominant, where the flood and ebb are of equal magnitude. The process of land claim, and associated activity such as dredging, can modify this flood–ebb cycle. It becomes easy to see how, if this flood–ebb relationship of a system is upset, the movement of sediment into and out of an estuary can be altered, and in an extreme case the dominance can be switched from flood to ebb (or vice versa), with the result that a once accreting estuary can become an erosional one or a once erosional system can become accretional.

While it may be considered beneficial for an eroding estuary to become net accretional, there are situations in which this can cause problems, such as near important harbours, where accretion of sediment may induce the need for increased dredging. Similarly, the increase deposition of fine sediments can induce the onset of marsh formation and the potential switch from sand to mud domination. The obvious problem here is where a resort at the mouth of an estuary begins to see its sandy beaches overrun by marsh.

Examples of such variation in tidal dominance are given by Orford (1988) and Orford and Carter (1985), and summarised by Carter (1988), using sites in eastern Ireland as examples. In this work, the environmental impacts of land claim are demonstrated for Rosslare Bay, where beach loss and spit depletion had been blamed on the impact of construction in Rosslare Harbour and interference in longshore drift. Research demonstrated, however, that the impedance of drift was insufficient to account for the magnitude of change observed, although major land claim in the adjacent estuary was shown to have significantly altered the tidal regime, with resulting modifications in sediment transport. This change in sediment regime caused sediment starvation in the spit region because less material was being brought out of the estuary on the ebb tide, owing to its becoming less dominant. Because of this starvation, waves began to rework sediment stored in the spit. This, therefore, highlights a situation where, owing to land claim, sediment dynamics were altered, shifting the environment into disequilibrium. As a direct result of the reduction in the strength of ebb currents caused by changes in the channel due to land claim, the lack of sediment caused the spit area to become net erosional (that is, more sediment was removed than was added), as the sea started to rework one of its sediment stores.

This example is one of many similar studies. Carter (1988) cites a further example, also from Ireland, where land claim in Malahide (Dublin Bay) resulted in the reduction of the tidal prism by 2 million cubic metres. Reduced ebb tide velocities caused erosion at the estuary mouth because, following land claim, the estuary switched from a situation of net seaward sediment movement to net landward. Hence, sediment was lost from the estuary mouth and transported up–estuary. Furthermore, reduction in ebb velocities has allowed longshore drift to encroach on the mouth by the formation of a spit, further modifying the estuary.

Loss of coastal habitats

Land claim is also responsible for the loss of considerable areas of natural habitat. Table 3.2 details the habitats lost from land claim in different areas of the coast. In effect, the process of land claim can lead to loss in any of the habitats which can be seen around any of the world's coastlines. As an illustration, the loss experienced around the coastline of England over the past twenty years, and predicted for the next twenty years (Pye and French, 1993a), is shown in Box 3.1.

The loss of coastal habitats can occur in three ways:

1   Habitat can be lost by land claim and land alteration, as for example the conversion of salt marsh to arable land or to industrial usage, or the planting of dunes for golf courses.

TABLE 3.2   The loss of coastal habitats as a result of land claim in a range of coastal environments

Coastal area	Habitats lost	Examples
Sand dunes	Vegetation communities	Planting of artificial grasses – golf courses
		Afforestation – coniferous
	Bog areas	Draining/lowering of water table
	Sand (infauna)	Extraction – sand removal
Saltmarsh	Vegetation communities	Agriculture, industry, waste disposal
	Reed swamps	Agriculture, industry, waste disposal
	Freshwater transitions	Agriculture, industry, waste disposal
Intertidal flats	Mudflats	Dredging, port development, extraction
	Sandflats	Dredging, port development, extraction
Shingle	Vegetation communities	Extraction, development
	Saline lagoons	Barrier modification, freshwater input
	Shingle	Starvation due to groynes
Cliffs	Vertical face	Loss of nest sites due to sea walls
	Cliff-top grassland	Development

Note: Because all habitats can also be lost by sea defence works, and by building, these two processes are not listed under each environment.

BOX 3.1 Loss and predicted loss of intertidal habitats along the coast of England

The English coastline stretches for around 10,000 km. Estimates of this length vary considerably depending on how the coast is defined. The figure of 10,000 km is that quoted by the Coastwatch survey, a survey monitored by the Joint Nature Conservation Committee (JNCC), and includes in its total any area under tidal influence; that is, lengths of all tidal rivers, etc. Other estimates quote figures as low as 2,900 km (Herlihy, 1982), referring only to open coast-line. Although this distinction is not critical from the point of view of this book, it is important to bear it in mind when reading literature which quotes habitats as percentages of total coast length.

The English coast is extremely varied in respect of its landforms, a reflection partly of its geology, its range of estuaries and coastal types, and also of its differing exposure to wind and wave processes. A major study undertaken for English Nature (Pye and French, 1993a) identified the result of human impact on all the major coastal morphological types. The study identified historical loss, and predicted future loss, with the aim of allowing English Nature to undertake the sorts of habitat creation projects described in this chapter in order to offset any losses, and thus to maintain levels of habitat at those of the early 1990s.

The causes of loss, and future threats are many and varied and are discussed in greater detail within the main text. However, the amount of habitat lost is significant, and represents, in percentage terms, the direct effects of human intervention in coastal processes. The table summarises the predicted future losses into the first decade of the next century.

Habitat	Present area (ha) (1992)	Predicted loss (ha) (1992–2010)	
Sand dunes	11,897	240	(2%)
Saltmarsh	32,462	2,750	(8.5%)
Intertidal flats	233,361	10,000	(4.3%)
Shingle landforms	12,376	200	(1.6%)
Saline lagoons	1,215	120	(9.9%)
Soft cliffs[a]	256 km	10 km	(3.9%)
Cliff grassland[b]	1,895	150	(7.9%)
Coastal heath	462	50	(10.8%)

Notes:
[a] 'Soft cliffs' refer to cliffs composed of unconsolidated (drift) material, or known unstable geological formations, which are not protected by any form of defence. The predicted loss refers to areas where defences will have to be built, in view of retreat rates over the past twenty years, in order to protect land of high value which will become threatened over the next twenty years.
[b] 'Cliff grassland' refers to areas of grassland on top of cliffs under the maritime influence.

As we have already mentioned, if the sea is prevented from flooding areas of marsh, the ecological structure of the area is naturally going to shift from one which is salt tolerant to one which is freshwater tolerant. The conversion of marshes to farmland will completely remove all natural vegetation, as this tends to be replaced by crops, or other grassland planted to maximise meat or milk yields in livestock. Complete loss will also be experienced when land is claimed for development, in which case marsh vegetation is often replaced by brick and concrete.

It is not necessary to remove habitats completely to cause vegetation loss. Changing land use can modify habitats, perhaps by increasing drainage, by supplementary planting with more favoured species, or by the promotion of species as a reflection of human action, such as the increased dominance of trample-hardy species as opposed to susceptible species, or by the grazing of animals.

While the modification of species can occur in any coastal habitat, perhaps the classic example of this is management of dunes for recreation, and particularly golf. There are typically two schools of thought with regard to this activity. The more purist view is that by the very action of converting dunes to a golf course, you are shifting the natural ecosystem balance to one of unnatural grassland, and turf management, and, as a result, modifying the fauna of the area. The second, and somewhat opposing, view is that the use of dunes for golf courses is perhaps a best compromise situation because it prevents total loss from other forms of development; it prevents the trampling problems and subsequent erosion of dunes experienced along many coasts; and it also provides the habitat with some degree of management protection. In addition, correctly maintained rough areas can provide homes for natural dune species, and eventually, if abandoned, the golf course would naturally revert to a semi-natural state. Thus some people would argue that far from the habitat being lost from the natural coastal system, it is in fact being borrowed, and could, without too much difficulty, be returned at a future date. While this may satisfy the ecological argument, it does not satisfy the process argument, because by conversion to golf courses, the sediment within the dunes is effectively removed from the sediment budget since new grass planting causes greater root bonding, and the increased land value also suggests that defences may become more likely.

2   The alteration of intertidal areas, without their being claimed for freshwater environments but nevertheless being modified in a significant way, such as by the building of ports or marinas, can also lead to significant loss and hydrological modification. In this case, the result is often that intertidal areas are converted to subtidal areas, either by dredging or by the artificial retention of water to maintain a condition of continuous high tide. Such a situation is that of port or marina construction. In the UK, the greatest single threat to intertidal areas is the planned construction of marinas, largely for private boats and yachts (Pye and French, 1993a). The survey by Pye and French (1993a), carried out in 1993 as part of English Nature's 'Living Coastline' initiative, showed that within England alone, there were fifty-nine proposals for marina development in estuaries, of which thirteen were in Hampshire (on the south coast), and two-thirds were within the south-east. This fact, although largely reflecting the population distribution and the reputed relative prosperity

within this corner of the country, also coincides with the area of greatest threat from sea level rise, an area where sea defences are known to be inadequate, and also an area where any loss of intertidal areas can have important implications for estuarine hydrology.

3   Finally, habitat can be lost as a result of the building of sea defences; that is, by cutting off sediment supply from eroding cliffs, with knock-on effects on beaches; or by preventing the landward migration of habitats during periods of sea level rise (coastal squeeze). In effect, sea defences fix a habitat in space, which means that if external factors, such as sea level, change, the habitat has to change form, rather than migrate. Migration will allow habitats to maintain the same form, but occupy a different space (Figure 3.3).

The need for dredging and spoil disposal

The modifications to the remaining intertidal area caused by land claim require that if access to ports is to be maintained, the channel has to be artificially maintained. This maintenance often takes the form of dredging (the removal of sediment from an area in which it causes an obstruction).

Dredging, however, does not occur only in response to land claim, but can apply equally to other aspects of channel maintenance. Irrespective of cause, its impacts can be significant. In effect, dredging is a way to deepen a channel that impacts on the rest of the intertidal area and sets in motion a series of environmental modifications.

First, by deepening channels, it is possible to modify the ratio of intertidal to subtidal areas, and thus alter the area of exposed mud available for feeding. Second, and more significantly, dredging alters the intertidal profile. Generally, a mudflat will slope towards the deepest part of the channel. When this channel is deepened by dredging, this slope will become unstable and in disequilibrium with the environment. It will therefore begin to adjust to its new base levels (Figure 3.4). At ebb tide, water drains from the marsh and high flats towards the low-tide channel. The steeper profile will provide a steeper gradient for this water to flow over, and there is, therefore, a greater chance of creek incision and gulley erosion into the mudflat.

Such processes occur as a result of the adjustment of mudflats to a new base level, and also because of the incision of creeks into the mudflats. This, therefore, poses a question: what happens to the volumes of

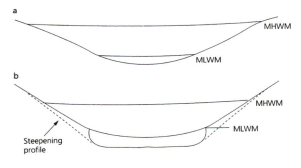

FIGURE 3.4 Impact of dredging on an estuarine deep water channel: (a) before dredging; (b) after dredging

MHWM = mean high water mark; MLWM = mean low water mark

sediment involved (the dashed areas in Figure 3.4)? Increased sediment erosion in one part of the estuary will provide increased sediment for deposition elsewhere. While this may work out favourably if deposition occurs in an area of need such as an eroding mudflat or salt marsh, it is equally possible that the eroded material will be transferred into the dredged channel, thus necessitating further dredging. Such redredging of a channel would be expected, however, because dredging causes a system to become unnatural and, hence, unstable. As soon as it becomes unstable, the system starts to revert to a stable form, with readjustment often including the natural infilling of the channel.

Another major problem with dredging is what to do with the removed material. Large quantities of muddy sediments can be produced by dredging. In some cases, where dredging is occurring around ports, the contamination levels in the dredged material are so high that the only safe environmental option is to treat the dredgings as contaminated waste, and to process accordingly.

In other cases, one of two options can be employed. Traditionally, the material is taken offshore and dumped in a licensed dump site. In process terms, this represents a transfer of sediment out of the nearshore system into a largely inactive sediment sink. Bearing in mind that we are regarding systems as being in a delicate balance between energy, sediment input, and sediment outputs, the loss to a system of large quantities of sediment can produce major deficits to the local sediment budget. A further problem with transfer to offshore dump sites is that large volumes of sediment are dumped into the sea, increasing turbidity and blanketing the bottom with a layer of sediments. This process can impact upon many aquatic life forms, both nektonic and benthonic.

Increasingly, a second approach is becoming more widely practised. In the second part of this chapter, we will look in detail at alternative methods of protecting the coast, and the impacts that these methods have on the coastal environment. We will see that 'soft' methods – that is, the creation of coastal habitats as a means of defence – are becoming more widely used. If dredgings are of suitable grade (grain size) and sufficiently free from contamination, then a good use for them is in the creation of such coastal habitats, whether by their use in beach feeding or in mudflat accretion. By carrying out such a process, the creation of habitats helps defend coasts but, perhaps more significantly from the sediment budget point of view, keeps the sediment within the sediment system.

Coastal protection

So far, we have discussed coastal defence as something which occurs alongside land claim, to protect the new land areas from inundation by the sea. Typically, when we think of this form of defence, we usually mean earth embankments along the shores of estuaries (Plate 3.2). The subject of coastal defence, however, is considerably more diverse than this when we come to the open coastline because it also represents one of the main ways in which humans can significantly alter coastal processes.

PLATE 3.2   Typical earth embankment protecting land claim from tidal inundation
(Old Hall Marshes, Tollesbury, Essex)

At regular intervals the media cover major coastal events, such as cliff falls
which threaten people's homes or businesses, and publicise the demands of those
involved to have their properties protected. Such publicity generally revolves around
cliffs and is concentrated in specific geographic areas. Cliff recession has been well
documented historically, although, despite such extensive documentation, little
appears to have been successful in terms of effective defence construction. Within the
UK, for example, it is possible to identify areas where cliff instability has affected
human populations, and where there are no defence measures in place. It can be seen
from Figure 3.5 that in England the greatest concentrations occur around the north-
east coast, north Norfolk and Suffolk, and the coast of Dorset. In recent years,
well-publicised media events have included the loss of the Holbeck Hall Hotel,
Scarborough (Clements, 1993), and the rapid erosion of the Holderness coast and the
associated loss of farmland. Historically there have been the loss of villages (Prestage,
1990), the loss of houses and land along the Dorset coast (Case, 1984), and the
historical loss of houses along the Suffolk coast (Parker, 1980). Although these are
examples of the erosion of the coast affecting the human population, it is, as we shall
see later, the very fact that humans have not interfered with these coastlines that has
saved other areas along the coast from not experiencing similar problems.

Sea defences are often regarded by people as being the ultimate saviour when
areas of coast or properties are at risk from tidal flooding. In many cases, this may
well be so, but if a sea defence scheme is not carried out with due consideration
for its full effects on the environment, then additional problems may well arise.

FIGURE 3.5   Areas of unprotected eroding cliffs in England
Source: Modified from Pye and French, 1993a

These new problems, however, need not necessarily occur at the point of defence construction, as the newly built defences merely cut off the sediment source. It is the areas where this sediment would have been deposited which will feel the effects of its loss, as there will no longer be the necessary supply of material to maintain beaches, dunes, or marshes. As a result, the defence of an urban area on top of some cliffs may well protect that area, but other settlements down the coast, which had not previously had any problem with erosion, may well find that new problems have arisen because the defence of the cliffs has stopped the supply of sediment which had previously gone to supply the beaches down-drift (see Box 3.2 for a specific example). The significance of eroding cliffs to sediment input can be seen clearly in the following example from Thompson Island, Massachusetts (Jones *et al.*, 1993). Here, cliffs have a geological composition similar to those referred to earlier at Holderness, eastern England. At Thompson Island, average cliff recession rates range from 0.4 to 0.6 m yr^{-1}, and this, coupled with cliff height and length data, suggests an annual input to the local sediment budget from these eroding cliffs of around 6,500 m^3 yr^{-1} (calculation based on the total input between 1939 and 1977). Such a volume of sediment is significant in the context of longshore sediment movement. Were these cliffs to be defended, then the local sediment budget would be deficient by this amount, and would attempt to compensate by eroding material from elsewhere, such as beaches. At Thompson Island, the beaches are also eroding, contributing a further 5,500 m^3 yr^{-1}, so this source may not be sufficient, possibly indicating that other sediment stores further down-drift may become erosional.

Other work by Correia *et al.* (1996) has demonstrated a similar phenomenon occurring along the Algarve coast of Portugal. Here, the rates of cliff recession have recently been estimated at 3.0 m yr^{-1}, having increased from 0.7 m yr^{-1} following sea defence construction further along the coast, thus illustrating how the sea will compensate for reductions in sediment input once one source is defended. Figure 3.6 details a hypothetical situation which may arise when an area of cliff line is protected, and also helps to explain the Portuguese coast case. In Figure 3.6a, erosion of the cliffs provides a sediment supply which is moved along the coast by longshore drift to enable the build-up of beaches down-drift. As a result, the loss of land at the cliffs is balanced by the build-up of beaches and a spit down-drift. In economic terms, this process provides a beach for tourism, as well as the spit, which shelters the estuary mouth and provides sufficient protection to allow the development of a port there.

After a period of time, a settlement on top of the cliff becomes threatened by the erosion of the cliff, and it is decided that some form of coastal protection is necessary. A sea wall is built along the base of the cliff to stop any further loss, and to protect the cliff-top settlement. These engineering works prove successful, and erosion at the cliff stops, and the settlement is saved. However, looking back at the relation in Chapter 2:

$$\text{Wave energy} = \text{erosion} + \text{transport}$$

without the ability to erode the cliffs, the wave energy transports sediment from the beaches in front of the now defended cliff using the surplus energy no longer being used for cliff erosion. As a result, this beach in front of the cliff is eroded until all the beach material has been moved along the coast, meaning that the waves hitting the coast now have a surplus of energy which is, in effect, transferred down-drift, and so the focus of erosion is moved laterally, with the result that in Figure 3.6b, erosion has started down-drift of the cliffs at a previously accreting beach, posing the same problem as before but shifting it laterally to another settlement or land use. In time, this new centre of erosion starts to threaten additional urban areas and so it becomes necessary to defend this new part of the coast, and in time the same process will happen again, and the focus of erosion will shift laterally down-drift (Figure 3.6c).

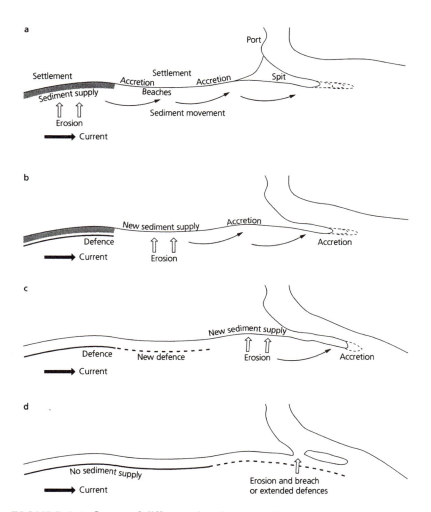

FIGURE 3.6 Impact of cliff protection along a coastline

This process will continue along the coast until the end of the coastal cell is reached. In our example, this point occurs at the spit (Figure 3.6d), where erosion means that the estuary becomes more exposed to wave energy, and also the operations of the port become threatened.

Ultimately, then, the initial decision to defend one section of the coast, where this section represents the primary sediment source, has important knock-on implications for the rest of that coastline. The final outcome is a coastline which is almost completely defended, to which access may be difficult, and which has a greatly reduced potential for economic activity, such as tourism or port activity. Had the cliffs not been defended in the first place, the cliff-top settlement might have fallen into the sea, but the rest of the coast would have remained natural. Alternatively, soft engineering techniques, such as beach renourishment, could have been used to reduce the problem by artificially adding sediment to the base of the cliff, allowing the constant reworking of this material instead of shifting the focus of erosion laterally.

These problems are not restricted to cliffed coasts. The preservation of beaches, too, especially in an area of tourism, is important from economic perspectives. The utilisation of defence techniques for beaches, however, also has serious implications for beaches down-drift. Box 3.2 details a case study in which these very problems are exemplified.

Box 3.2 highlights the problem of hard defences and beach sediment supply and loss. A study by Hall and Pilkey (1991), looking at the widths of high beaches (between neap and spring tide high water mark along sandy beaches in New Jersey), highlighted the impact of hard defences on beach width reduction. This reduction in width occurs because the tide comes in further when beach levels are lower. Of all the beaches studied, 51 per cent of those backed by sea walls had no high beach; that is, the tide came in to the wall. Where no defences were present, all beaches had at least 1 m. Where groynes were present, beaches were generally absent from the down-drift side, but present up-drift where sediment is trapped. Summarising their findings, Hall and Pilkey indicate that for beaches backed by walls, the mean width is 9.0 m (range 65.5–0). For coasts with groynes, this average width increases to 18.5 m (range 39.0–7.0), while those beaches with no defences averaged 35 m (range 454–6). These results appear to suggest that defended coasts have, on average, narrower beach widths or, put another way, experience greater beach lowering.

De Ruig and Louisse (1991) look at the sediment budget problem in a wider context. On the Dutch coast, which is largely claimed land, much of the sediment supply comes from the North Sea. Circulation patterns in the North Sea tend to be anticlockwise, with one of the prime sources being the Holderness cliffs of eastern England. This study has indicated a net sediment input to this part of the Dutch coast of the order of 0.4 million $m^3 yr^{-1}$. This would suggest a significant and constant sediment supply, which could become deficient if Holderness were ever to be defended.

These problems and the examples cited bring us on to what is perhaps the most contentious issue in coastal management at present. Should a coastline be protected from erosion, even when that coastline is developed and provides homes and employment for people? Or, to ask the same question another way, should coastal authorities allow people's homes to fall into the sea when defending them may lead to additional, and more severe, problems elsewhere along the coastline?

Many authorities have differing views on this subject, but we live in an environment of increasing use of soft engineering techniques, and one in which there is pressure to protect the naturalness of the coastline; to avoid expense; and to avoid having unsightly concrete sea walls, with the implications for future finance of eventually having to build new walls. For these reasons, the growing trend is to allow erosion to continue in areas of low land value, even if this means that individual houses or businesses should fall into the sea.

BOX 3.2 Coastal protection and its impacts along the Fylde coast, north-west England

The Fylde coast is a wind- and wave-dominated coast located in the north-west of England. Geographically, it comprises a north–south-trending coast stretching southwards from Morecambe Bay to the Ribble estuary. Historically, this coast has undergone extensive development, resulting in an almost continuous sprawl of urban/commercial development, including such resort areas as Blackpool and Lytham St Annes. This development has led to increased protection for almost the entire coastline, owing to the increasing value of the hinterland.

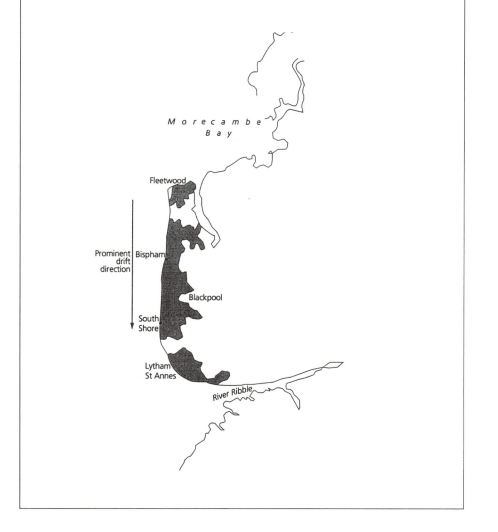

Morphologically, the coastline is sand dominated, consisting of wide sandy beaches with gravel increasing in abundance northwards, and an extensive active dune complex to the south of Blackpool and into Lytham. Traditionally, sediment sources included soft glacial cliffs which occur along this coastline, and also from the predominant north–south drift direction. The direction of drift is further complicated by periodic reversals due to prevailing wind directions and the sheltering effect of Ireland. Although significant, these transport reversals are not critical in the demonstration of the impact of defences on beaches and sediment movement, and so will not be discussed in detail to avoid unnecessary complication.

Increased coastal defence works removed the soft cliffs from the sediment budget of the area, significantly reducing the sediment input to the coast. Following this, the only sediment source was from southerly drift from the north. The problem of reversals in sediment movement and the cessation of inputs from eroding cliffs has meant that beaches in the area started to experience sediment losses and beach lowering. The response of the two local authorities along this coast was to install groynes to retard the southerly drift of sediment.

Although these were effective in retaining sediment, the impact on the coast further south was significant, with beach losses and detrimental impacts on the dunes south of Blackpool. As a result these groynes were removed from the coast from Bispham southwards, effectively freeing up this section of coast and allowing unimpeded sediment movement southwards from this point. The groynes to the north, however, were left in place, thus splitting the coast into two sections, with different defence methodologies. The changes in defence policy occur at a local authority boundary, and serve to reinforce the issues discussed in Chapter 7, where two groups control different parts of a coastline which is acting as a single sedimentary unit.

The effects of this human interference have been marked. From the north, the groynes around Fleetwood are full and sediment is passing down-drift. To the south, the groynes are only partly filled, although enough sediment is being retained to provide adequate protection for the defences present. At Bispham, where groynes have been removed, beach levels are low, owing to continuous sediment movement south but, because of the groynes, little replenishment from the north. The unprotected sandy beaches of the Bispham and Blackpool areas are, in effect, acting as the sediment source for southerly drift, and are therefore experiencing a net loss of sediment and subsequent beach lowering. This problem is causing great concern for the local authorities at Blackpool, a problem which is being further aggravated by the vertical nature of the sea defences and the subsequent high reflectivity of the structures.

Blackpool's loss of sediment has to be gained by some other area. Sediment movement down-drift occurs until the end of the defences at South Gate, when the significance of the wind becomes important in that large quantities of sediment are transferred into the sand dunes. At this point, the beach levels are good and there is abundant sediment exchange between the beach and dunes. Further south, the drift is held up at the estuary of the Ribble. Here, large sandbanks have developed, representing large stores of sand. The movement of this sand across the mouth of the Ribble is relatively slow, and can occur only when storms force the channels to move, allowing sand to shift its position relative to the estuary's ebb flow, and subsequently continue its southerly migration down the coast to Southport and the Ainsdale dunes.

To summarise the problems of this coast, southerly movement from Fleetwood is slow and sporadic, dependent on the degree of groyne infill. Further south, the lack of groynes means that sediment movement exceeds replenishment, resulting in beach lowering and net sediment loss. Further south, large quantities are transferred into the dunes and into sand-bars. Indeed, between South Shore and Lytham, the build-up of sand is sufficient to support extractive industry, with large quantities of sand being removed commercially.

In considering the question of whether to defend or not, the coastal engineer needs to consider two questions:

1   What is the value of the land which is being threatened; that is, what are its uses (arable, urban, industrial), and what environmental threats might be posed if it were allowed to erode (contaminated land, threats to rare habitats, loss of housing/industry, etc.)?
2   How much would it cost adequately to defend the area in question – the cost of building concrete walls, the creation of new habitats, etc., as offset against the value of the land involved?

In addition, there also needs to be consideration of other socio-economic factors, which may tip the balance of the decision, such as the effects on the local economy of allowing a local farm to fall into the sea, the impact on a village of allowing an access road to be lost, and the cost of replacing it (Table 3.3). When all parts of the equation have been considered, it comes down to a question of which of the above points (defence or no defence) will have the greater cost, and what, therefore is the most cost-effective option to undertake. What is usually done is essentially a cost–benefit analysis into whether the land or properties which will benefit from protection are worth it in a financial sense; that is, is it worth building defences at a cost of several million pounds to protect a group of four houses valued at around £100,000 each? This, of course, neglects the personal aspect, the stress and feelings of the people who live in those houses and whose lives would be affected by the ultimate decision, and on which the placement of cost is impossible. As an example, in the late 1980s, Rother District Council (near Hastings, Sussex, in south-east England) rejected coastal

TABLE 3.3 Practical considerations in the decision as to whether or not to construct sea defences

Costs weighted for defence	Costs weighted against defence
• Protection of land of high value (industry, ports, etc.)	• Implication for coasts down–drift (cutting off sediment supply)
• Income from arable land	• Costs of building defences
• Jobs at stake through land loss	• Loss of intertidal habitats
• Number of houses to be lost through erosion if no defences built	• Effects on tourism
• Effect on local economy	• Low value of hinterland (arable)
• Replacement of infrastructure	• Impact on the function of the coastal cell
• Effects of planning blight	
• Effects of flooding	

defence schemes ranging in value between £1.8 million and £9 million on the grounds that all that would be protected would be a few houses (Meyer, 1991). In this case the value of the houses, and the land in general, was not considered of sufficient value to merit any protection by defence construction. Following on from this, however, another scheme was put forward and awarded funding. This scheme involved a soft engineering method, and involved the artificial building up of the beach in front of the cliffs, reducing wave erosion in this way. This process, known as beach feeding, involved the placement of 250,000 m^3 of shingle to act as a protective beach in front of the problem cliffs, and cost considerably less than the original schemes, involving the construction of hard defences, would have done. The cost–benefit analysis meant that this proved to be financially viable. A full discussion of the procedures used in such decision-making can be found in Chapter 7.

It is quite clear that the decision to defend a stretch of coast is a complex one, and one which involves many aspects. The general methodology behind such decisions needs to take all such concerns into account.

Impacts of management techniques on wave-dominated coastlines

The natural processes on wave-dominated coastlines involve the erosion, transport, and deposition of sediment to form beaches and other sedimentary landforms. On inhabited coasts, these processes can become interfered with, with the result that management problems can occur. Each form of coastal defence used will produce its own set of effects on sediment movement and erosion, and its own set of environmental problems. The sorts of issues raised include:

1   excessive material drawn offshore and stored below normal (non-storm) wave base by shore-normal currents producing beach lowering and increased erosion,

caused by a deficit of material in the swash (that is, lack of material coming from up-drift owing to other defence works and sediment retention methods) and, as a result, excess wave energy leading to greater removal of material on the backwash;

2   beach degradation and down-drift sediment transfer by longshore movement due to non-replenishment of source area as a result of up-drift defence and sediment retention measures, therefore leading to increased erosion rates which gradually prograde down-drift along the coast;

3   enhanced erosion in bays and at headlands due to concentration of waves from refraction around breakwaters or walls;

4   the encroachment of spits, etc. on river mouths, and subsequent navigation problems.

All the above issues can be summarised as occurring in one of three prime areas: sediment supply, coastal morphology, or current directions. These are the three areas in which studies of coastal problems need to be aimed. There is, however, an alternative to these three approaches, and that is not to tackle problems 1–4 above, but to look instead at the root cause; that is, the waves and tides. This then outlines the initial approach strategy in dealing with protection along wave-dominated coasts: does one tackle the problem from the sea (i.e. the waves) or from the land (i.e. the coast, sediment, etc.)? Each is a possibility, and each has its own benefits and problems.

The approach taken will govern the form of defence strategy used. Each method of coastal defence is outlined below, along with an indication of environmental impacts and examples where the technique has been used.

Offshore breakwaters – stopping the waves in the first place

The theory goes that if you can stop large waves hitting the coast in the first place, then problems of erosion and unwanted modifications to coastal morphology become less significant, ideally to the stage where they are no longer a management problem. The use of offshore breakwaters provides a suitable method by which wave energy can be dissipated further seaward than would otherwise happen and, hence, not occur at the beach or cliff. There are two ways in which this can be achieved:

1   increasing the frictional drag of the sea bed by either decreasing water depths or increasing bed roughness (see Chapter 2); or

2   physically stopping the waves by means of a permeable barrier, such as piles of rocks through which waves have to pass, or by using an impermeable barrier which prevents shoreward movement of the wave altogether.

Irrespective of the design method chosen, it is not possible fully to remove the potential of waves reaching the coast, as there are still problems of waves reaching the shore by processes such as refraction and diffraction around the ends of the structures. This, however, can be used to the coastal manager's advantage. We have seen how sediment circulation cells often occur in conjunction with current activity, and if this effect is used in the construction of an offshore breakwater, then results can be favourable. If such wave refraction is utilised to force a sediment transport pathway past the breakwater, then in the lee of that breakwater, where conditions are quieter, sediment will be deposited, resulting in a sand bar eventually linking the shore to the breakwater, in the form of a tombolo. This process may be desirable at the specific coastal locations where offshore structures are built, but there is the risk that sediment entrapment may be so effective that problems of sediment starvation occur down-drift of the breakwaters, thus merely transferring the original problem laterally.

When using such features, design is important, as enough material is needed in the construction physically to do the job – what you do not want is to have material which is too light, and so is moved by waves, as all this does is provide ammunition for the waves to throw against the coast in times of storms. It must be borne in mind that there have been recorded sightings of waves 400 m long and 11 m high that have lifted 40 tonne concrete tetrapods.

ENVIRONMENTAL IMPACTS

The most obvious impact of offshore structures is to reduce wave energy at the shore. As this is the intended effect, it cannot be considered as detrimental in its own right, although knock-on effects may lead to problems. Other environmental implications are:

1   The build-up of sediment in the lee may be beneficial for beach construction, amenity use, and for a tourist beach, but it will also cover shallow benthic ecosystems, swamping them in sediment, effectively replacing a wave-cut platform containing rock communities with fine sediment. Such actions can cause great ecological damage and lead to the loss of whole communities, although new, but different, communities will replace them.

2   The reduction in wave energy will reduce erosion of the coast behind the barrier. Effectively, a once-eroding coast will become a depositional one. Up-drift, the movement of sediment will continue along the coast, producing sediment deposition immediately up-drift and at the site of the structure. In addition, down-drift the presence of these structures will cause an interruption to this longshore sediment transport and may well decrease deposition in these areas, resulting in beach sediment removal and the subsequent initiation of a new erosion problem. In essence, the barrier will stop a prime source of sediment, which may have additional implications for the stability of adjacent coasts.

3   Decreased wave activity will reduce the wave splash onto supratidal environments, meaning that dune communities, heaths, and maritime grass-lands can lose their maritime influence, and so ecologically revert to terrestrial vegetation.

4   The foreshore will become partly isolated from the active coastal environment, which in some cases can lead to a reduction in the amount of sediment available for seasonal fluctuations in beach levels; that is, sand dunes supply sediment to the foreshore during the stormy winter months to modify the beach profile, while storing it during the summer.

5   Visible intrusion and aesthetic effects must be considered, and also the fact that a hugh offshore breakwater reduces the accessibility of the coast to ships and pleasure craft.

Not all impacts are negative, however. New conditions to landward of a breakwater can favour new communities, while the breakwater itself can act as a reef structure, proving a completely new habitat.

Despite these problems, offshore breakwaters are becoming increasingly common as a means of coastal protection. Many have been used effectively to trap sediment and build up beach levels along coasts. In England, such schemes are operating in Norfolk, Suffolk, Essex, Sussex, and Devon. In Sussex, a series of break-waters has been constructed to facilitate an increase in beach levels and reduce the flood risk for coastal settlements (Pethick and Burd, 1993). Along the North Sea coast of Denmark (Moller, 1992), a coastal defence scheme at Jylland has employed a series of strategies, one of which is offshore breakwaters. This is a highly dynamic coast with large longshore sediment movement, but is also one which experiences erosion. Breakwaters of basalt blocks have successfully been used to form a series of tombolos, effectively connecting the breakwater to the coast.

Not all schemes are as successful, however. Fox (1997) reports on an example from Sidmouth in Devon, south-west England, where during the first winter of operation of the breakwater scheme, beach levels have dropped dramatically owing to storm waves which occurred at angles not covered by the breakwaters. This may, however, rectify itself as sediment is brought back onshore as the area adjusts to its new equilibrium state.

Artificial bays

Natural coastal embayments aid coastal protection in two ways. First, their morphology results in a series of small sub-cells which trap coarse sediment leading to localised beach development. Second, bays effectively lengthen the coastline without any increase in wave energy, and so reduce wave energy per unit length of shore. Both these processes can be used as part of coastal defence strategies.

On embayed coastlines, alternating hard and soft rocks cause a series of headlands to develop, with softer rocks eroded to form the bays. On cliffed coasts, where problems of wave erosion occur, it is possible to produce these hard points artificially, and thus encourage the formation of bays. The main problem which does occur is that natural bays are associated with characteristic offshore bathymetry, which results in the wave refraction patterns we have already seen (Chapter 2). Artificial bays do not have such bathymetry, and so they tend to be less stable. By artificially manufacturing embayments, it is possible to promote coastal deposition, and reduce wave energy. The final bay shape and size depend on the wave regime and the direction of wave approach. The ability of the bay to trap and hold sediment is a function of the sediment grain size, as well as the length/depth ratio of the bay.

ENVIRONMENTAL IMPACTS

The development of an artificial shoreline has perhaps one of the greatest impacts possible for a stretch of coast because rather than afford protection to the coast in its natural form, it creates a completely new coastal topography. For that reason, this process should be considered only after all other possibilities have been studied. The new sedimentary environments will alter the local habitats for both flora and fauna. These effects are predictable. In contrast, the impact on the rest of the coast is less so because by creating an embayment, you are, in fact, creating a coastal sub-cell; that is, a means of trapping sediment which might normally travel further along the coast. In addition, there are also problems with accelerated erosion in the early stages of development; where is this eroded sediment going to end up, and exactly where is it going to be eroded from? This accelerated and partly unpredictable erosion means that this type of engineering is not suitable for all coasts, especially those where development of the hinterland is involved because the formation processes are largely unpredictable.

Because such schemes have such a great and unpredictable impact, their use has been somewhat restricted. Although I am not aware of any completed scheme, a proposed scheme does exist for the creation of an embayment at Dinas Dinlle, near Caernarfon, north Wales (Pethick and Burd, 1993). Here, conventional hard defences were not cost-effective, and so a scheme of hard points along the coast, with localised hard protection, has been devised to encourage erosion at specific points to create an embayment. The scheme is very complex in nature, and whether it is successful remains to be seen.

Beach recharge

Beaches are perhaps more efficient than anything else in dissipating wave energy, and hence protecting the hinterland. The energy dissipation is achieved by several methods: first, by providing mobile sediment which maximises the frictional area and allows grain movement, hence dissipating wave energy; second, by allowing free drainage through sediment but with decreased velocity, thus minimising the effects of backwash and, thus, offshore movement of sediment; and third, by permitting morphological changes to occur in response to wave conditions, thus increasing energy adsorption under different meteorological conditions.

The idea of beach feeding, as the name suggests, is to increase artificially the quantity of material on a beach which is experiencing sediment loss by erosion. As a result, the engineer artificially reproduces the conditions listed above. The situations where feeding occurs are largely where sediment losses have occurred owing to the construction of defences up–drift, leading to the prevention of longshore movement of material, with the result that the beach is starved of sediment.

Basically, the technique is simple. Material is either pumped onto the beach from a dredger offshore, or is moved onto the beach by dumper truck. However, it is not just a question of dumping large quantities of sediment onto the beach, as it is important to maintain beach profile and sediment grain size. There are two approaches to beach feeding: either estimate the profile and construct the beach to this, or overestimate the upper beach levels and allow the waves to draw material down to form a natural profile.

Source material is generally from offshore, but less occasionally land sources are used. Increasingly, dredgings are being used where they are sufficiently clean and of the correct size range. The important thing is to select the right grade of material, and similar rock types. If the recharge material is too coarse, then wave action will not be able to move it and the beach profile will remain the same throughout the year and not function as a suitable wave dissipator. In some cases, however, material is used which is slightly coarser than the natural sediment in order to increase the resistance of the beach to erosion. In such cases, an artificial profile needs to be established as waves will not be able to fashion their own. Newman (1976) illustrates this point for sand beaches, recommending a mean grain size for recharge material of 1.5 times that of the original. Conversely, if the recharge material is too fine, it will be preferentially removed from the beach, or will become washed into pore spaces between existing sediment and block the permeability of the beach to throughflow of water. McFarland et al. (1994) illustrate this problem for shingle replenishment in southern England.

Material is typically mixed with 10 per cent water to allow it to be pumped. This water will then drain off the beach, carrying the fines to seaward. Sometimes temporary bunds are used, or bulldozers to move material around. The exact technique will affect the eventual beach profile.

ENVIRONMENTAL IMPACTS

Beach recharge is perhaps one of the simplest methods of protecting the coast, in that it replaces what the waves have previously removed. However, it still acts against natural processes because the whole reason for doing the replenishment is that the beach is eroding, and so there are still impacts to be considered. These impacts can be divided into two groups: those caused during the pumping and beach build-up, and those which arise as a result of the new beach profile. Another review of impacts and the renourishment process is provided by Charlier and de Meyer (1995a).

The following are the first group of impacts, those arising from construction:

1   Run-off from the beach can cause scouring in the intertidal zone. We have said that 10 per cent water is added to allow pumping. This excess water has to drain and, if it becomes channelled, can erode.

2   Fine sediment can alter the ecology and the profile of the intertidal zone by swamping communities in sediment or increasing turbidity.

3   Loss of fines can cause the new beach material to become coarser than required and thus relatively immobile with respect to wave refashioning.

4   Use of heavy machinery can lead to compaction of sediment and hence reduce the pore space, and so reduce the throughflow from backwash. The result is increased surface flow with increased sediment removal and gullying.

5   Removal of material from the source site must be considered; that is, is the offshore source acting as a natural breakwater and thus protecting large sections of the coast? The removal of an offshore sediment bank or barrier which protected the coast from waves can lead to increased wave attack at the coast and make the original problem worse.

6   Pollutants or toxicity of the source material must be considered. Heavily polluted source material may damage environments in the intertidal or near-shore zone.

The second group of impacts, those occurring post-construction, are:

1   The wrong grain size can lead to rapid removal if the sediment is too fine, or an immobile beach form if too coarse. It is important for the beach to function as part of the sediment system of a particular environment, and so sediment needs to be of the same size distribution as that present in the area already.

2   Removal of fines can lead to a deficit of certain size classes on the beach, and so can reduce aeolian processes and produce dune starvation.

Despite the potential problems, beach feeding is the most widely used method for soft coastal protection, and is seen as a suitable way of disposing of unwanted dredgings. It is also a way to protect a site while maintaining a degree of naturalness, allowing or increasing amenity value, and could lead to the creation of new habitats such as saline lagoons. In addition to this, placing sediment on a beach involves no interruption to the longshore movement of sediment. Where beaches are starved of sediment owing to groynes up-drift, beach feeding can be seen as a way of artificially moving sediment past this obstruction, thus allowing natural processes to operate without the net loss of sediment, and subsequent erosion of material from beaches immediately down-drift of the defences. With reference to Box 3.2, a case study of the Fylde coast, readers could well consider such a solution to the problems of sediment loss from the Blackpool frontage, especially considering the commercial removal of this same sediment from the foreshore at Lytham for building.

With renourishment schemes being such a common method for coastal protection, there is a wide range of examples which can be cited. Table 3.4 lists some of these case studies in order to provide extended reading. In addition, Charlier and de Meyer (1995a) and Riddell and Young (1992) discuss general principles behind the methodology and practice of beach feeding, while Verhagen (1996) discusses the methodology with respect to the quantities of sediment needed and post-feeding losses. Estimates given by Verhagen indicate that only around 52 per cent of material

TABLE 3.4   Some examples of beach renourishment studies from around the world

Country	Author(s)	Location
Australia	Bird (1990)	Port Phillip Bay
Belgium	Charlier and de Meyer (1995b)	North Sea coast
Cuba	Schwartz et al. (1991)	Varadero Beach
Denmark	Moller (1990)	North Sea coast
France	Hallégouët and Guilcher (1990)	Brittany
	Anthony and Cohen (1995)	French Riviera coast
Germany	Kelletat (1992)	North Sea coast
	Eitner and Ragutzki (1994)	Norderney Island
Japan	Kadomatsu et al. (1991)	Toban Coast
	Koike (1990)	Tokyo Bay
New Zealand	de Lange and Healy (1990)	Tauranga Harbour
	Foster et al. (1996)	Mt Maunganui Beach
Portugal	Psuty et al. (1992)	Praia da Rocha
UK	Lelliott (1989)	Bournemouth
	Willmington (1983)	Bournemouth
USA	Walton and Purpura (1977)	(general)
	Campbell and Spadoni (1987)	Florida
	Giardino et al. (1987)	San Luis Beach, Texas
	Chill et al. (1989)	California
	Bocamazo (1991)	New Jersey
	Weggell and Sorenson (1991)	New Jersey
	Dixon and Pilkey (1991)	Gulf of Mexico
Various	Davison et al. (1992)	(various)

may be retained by the beach. This apparent loss of sediment, along with subsequent losses caused by longshore drift, highlights another important aspect of beach renourishment. Because beach feeding only compensates for sediment removed, the process needs to be repeated, the period of repeats being dependent on the volume of sediment used in the original fill and the rate of longshore transport. In Bournemouth, southern England, renourishment needs to occur around every eight to ten years. If renourishment is not repeated, then the original erosion problem may well return. Finkl (1996) highlights this as a problem along the Florida coast, where, owing to financial cutbacks, further funding for beach renourishment has been stopped, which could lead to the onset of new erosion problems. As Finkl points out, however, this is very much a false economy as renourishment costs equate to less than 0.4 per cent of the income from beach-related activities, and this excludes the higher cost of erosion prevention in the future.

Increasing sedimentation rates

Beach feeding involves the artificial supply of sediment to a beach by pumping material onshore artificially. It is also possible to increase natural sedimentation rates by preventing the loss of sediment from the foreshore by restraining its movement. This can be done either by constructing obstacles to longshore movement (i.e. groynes) or by stopping the removal of such large quantities in the backwash, such as by changing the properties of beach drainage.

By interrupting the longshore sediment movement, sediment can be contained on a beach where material is deficient. Groynes are the most commonly used device for this, with sediment being reworked within the groyne compartment to face the oncoming wave front, further increasing the potential for wave dissipation. When groynes become filled, material can then be transported into the next groyne, either by the burial of the groyne, or by transfer around the front.

If beach drainage is increased, water still flows back to the sea as a backwash, but carries less sediment with it. This is particularly important on sandy beaches, where most of the flow is surface flow. Problems also occur on shingle beaches, although in this case techniques are slightly different.

ENVIRONMENTAL IMPACTS

Interrupting sediment movement along any coast can cause serious problems down-drift with beach starvation, and the potential for increased erosion. Such problems have been mentioned elsewhere, and so need not be restated. Nevertheless, groynes have the potential to cause significant impacts on coastlines and, if not properly investigated, can lead to greater problems than they solve.

Increased drainage is perhaps a better alternative. By allowing water to flow more freely, through pipes or drainage channels, the problem of sediment loss can be reduced. Although not as significant as in the case of groynes, this increased sediment retention on beaches may lead to problems of deficit elsewhere along the coast because the problem still remains that less material is being transferred along the coast.

Examples of groynes and their impacts on coastal processes are even more numerous than those of renourishment. The case study of the Fylde coast (Box 3.2) illustrates problems which can arise. Other work (Livesey, 1997) highlights other problems which groynes can cause. Using newly installed fish-tail groynes along parts of the Morecambe frontage, north-west England, he illustrates how sediment accumulation has been so successful that large quantities of fine clays and silts have created a potential hazard for beach users, although as these consolidate, they are now becoming safer. In addition, sediment deposition has also inundated local shellfish beds.

Granja (1996) indicates how the installation of groynes along the north-west coast of Portugal has caused increased erosion in many places by cutting off sediment from parts of the coast. During storms in 1994/95, recession rates of local cliffs, devoid of their protective beaches owing to the groynes, ranged between 30 and

$60\,\mathrm{m}\,\mathrm{yr}^{-1}$. In other places along the same coasts, groynes have accelerated the erosion of dune complexes, threatening local housing developments.

Sea walls

The last of the techniques carried out on the upper beaches relates to physical protection of the upper shore margin. Sea walls are a solid barrier to waves and have been the most traditional method of coastal defence. Sea wall design varies greatly, and can serve either to reflect or to absorb wave energy. Reflective walls are generally vertical whereas absorptive walls are sloping, rough, and porous (see Carter, 1988, p. 444 for detailed descriptions).

Problems result if wave fronts and sea walls are misaligned. Reflection of waves in the cross-shore direction can cause standing waves to form in front of the defence, which sets up a series of scour cells that effectively lower the beach, and eventually undermine the sea wall. Because these structures form a physical barrier, they receive extremely high impacts. It is normal to expect around 90 per cent of sea walls to need some degree of maintenance within the first ten years, owing to either the wave battering or the material thrown at them by storm waves.

ENVIRONMENTAL IMPACTS

The most serious problems are wave reflection and foreshore scour. Foreshore lowering undermines the toe of the wall, which can collapse. There is also a flanking effect at the down-shore edge of the sea walls, where accelerated erosion occurs. Reports of such events are made by Hall and Pilkey (1991) and Griggs and Fulton-Bennett (1987), who cite inadequate construction, lack of awareness of process, and poor design as reasons for sea wall failure. Problems of beach lowering are discussed by Plant and Griggs (1992), who indicate that sea wall impacts on beach levels are caused by (a) alteration of swash velocity, direction and elevation which causes increased backwash velocity and increased seaward transfer of sediment; (b) higher beach water tables, causing less infiltration of backwash and further increased surface flow and seaward sediment transfer; and (c) greater variations in beach shape caused by (a) and (b), which lead to greater beach disequilibrium.

The main impact of sea walls is that they render the coast static; there can be no sea–hinterland interaction and whatever ends up behind the sea wall is no longer part of the marine area. As a result, sand dunes or cliffs can be sealed from marine influence. Ecologically, this can remove one of the main environmental factors from a habitat, namely the sea, and, as a result of protecting the land, also cut off sediments from reworking. Hence there also tend to be significant impacts on sediment budgets (refer to Figure 3.6).

Impacts of management techniques on tide-dominated coasts

In tide-dominated environments, energy levels are considerably lower and so large-scale defence structures are generally not required. Despite this, however, there are still problems with erosion and sediment movement which need human interference in order to protect development and general land usage.

As with wave-dominated environments, the best intertidal profile for sea defence is one which slopes gradually over a long distance, as opposed to one which slopes steeply over a short distance. With wave-dominated coasts, this typically takes the form of a wide beach. On tide-dominated coasts, owing to the lower energy conditions, the same theory applies but here the upper part of the intertidal zone (equivalent to the beach on a wave-dominated coast) is often mudflats, part of which may be vegetated by salt marsh. Despite this, the same logic applies.

A wide, shallow intertidal flat leading into a salt marsh is much better at dissipating energy than a narrow, steep profile. The added advantage here is that the marsh, covered with vegetation, has a high bed roughness and so its frictional drag is higher than that of a mud flat. Hence, the best defence for a piece of land is a mudflat and salt marsh.

In addition, we also have the advantage of vegetation cover providing a mechanism by which sediment can be trapped on the marsh surface by water passing between the stems and sediment becoming lodged. By this mechanism, marsh surfaces grow rapidly. In similar vein, once a marsh surface becomes grazed by live-stock, this vegetation cover is eaten down, and so its effectiveness as a sediment baffle is reduced. Similarly, while muddy sediment can be eroded, muddy sediment bound together by roots is much more resistant, and so the substrate becomes more stable.

Another point which should be borne in mind is how easy it is to erode the sediment. Sandy sediment is said to be non-cohesive; that is, it functions as individual grains which move independently of each other. In muds, grains act as a cohesive unit. Compare the effort needed to walk across a sandy beach and that needed to walk across a mudflat. Because of this, muddy sediments, even though they are finer grained, are harder to erode than sandy sediments; that is, it takes more energy to erode a mudflat than a sandflat.

There is a wide range of distinct management issues for tide-dominated coasts. The principal habitats of concern in tide-dominated environments are intertidal flats and salt marshes. In reality, salt marshes can be viewed as intertidal flats which have been vegetated. One of the key issues is that although wave energy on tide-dominated coasts is minimal, some does occur, and in such instances salt marshes are of prime importance in reducing it – and thus play key roles in the defence of such coasts. As an example of this importance, Table 3.5 shows the height of sea defences needed to protect land behind different salt marsh widths from flooding. It is clear that the wider the salt marsh, the lower the sea wall need be because the more wave energy is attenuated by the marsh surface. As a result, the cost of construction comes down dramatically from around £5,000 per metre for no salt marsh to around £400 per metre where there is a width of 80 m. Given that tide-dominated environments are

TABLE 3.5   The relationship between salt marsh width and the required height and cost
of the sea wall

Width of marsh (m)	Required height of sea wall (m)	Cost (m⁻¹)
80	3.0	£400
60	4.0	£500
30	5.0	£800
6	6.0	£1,500
0	12.0	£5,000

Source: Figures from the UK Ministry of Agriculture, Fisheries and Food

governed by the same principles as wave-dominated with respect to profiles and
energy absorbtion, it is also reasonable to suggest that changes in the intertidal zone
are going to cause similar changes in both environments.

Some of the key threats posing problems in tide-dominated environments are
detailed below.

1   *Increases in tidal range.* If tidal range increases, this will have two effects. First,
it will increase the chances of sea defence overtopping, and it will also cause modification to mudflat and salt marsh morphology and stability. The problem is caused
partly by increasing sea levels, but also by the construction of sea defences and
reclaiming of land (Figure 3.2). If a sea wall is built along a salt marsh, then it cuts
off areas which were once flooded by water. This water has to be accommodated
somewhere, and so can only be accommodated by an increase in water depth; hence,
high tides become higher. If tides become higher, then the volume of water moved
on each tidal cycle increases, and thus to compensate for this, current velocities
increase in order to move that increased volume of water into and out of the estuary
on each flood/ebb tide, the duration of which does not change.

2   *Changes in windiness.* The impact of wind and associated wave activity
has important implications for mudflat and salt marsh stability. Windiness in an
environment is an environmental factor which can change as part of the natural
atmospheric variation, producing periods of net storminess and increased calm. It can
be clearly seen in the Severn estuary, as well as many other estuaries, how increased
storminess can have serious impacts on salt marsh stability. In the Severn, for
example, it is possible to detect four distinct episodes of salt marsh development (Box
3.3). Each onset of new marsh growth follows a period of increased storm activity,
and associated erosion. Changes in windiness can also be associated with increased
frequency of meterological tides, and thus the problems of surge tides and overtopping of coastal defences. When these storm surges enter estuarine areas, average
tidal current velocities and bed shear stresses are increased, thus increasing the
potential for erosion.

3   *Natural changes in channels.* As with any channel, estuarine channels experience periodic shifts in position. Such channel movement can initiate a new set of erosion problems, as was the case at Silverdale salt marshes, Morecambe Bay, where the marshes started to erode once the Kent estuary channel moved towards the coast (Pringle, 1987). In contrast, the area on the opposite side of the Kent estuary, now some distance from the channel, is undergoing accretion and the rapid spread of *Spartina* sp.

4   *Dredging and navigation.* Estuaries, as we have seen, are ideal places in which to establish harbours and ports. Historically, this was not a problem because ships were relatively small and could sail into small channels. As ship size has increased, however, so ships need deeper channels in order to access ports which have not, by and large, shifted their location. Consequently, dredging has to be carried out in order to maintain access. Dredging has several effects which can threaten tidal environments.

(a)   By deepening the channels, it is possible to increase flow velocities and the amount of tidal discharge which can occur through the main channel, thus reducing flow and shear stresses across adjoining mud flats. As a result, it is possible to initiate accretion in the intertidal area. Such effects have been noted in the Dee, Mersey and Ribble in the UK.

(b)   To cause a progressive net movement of sediment from the intertidal flats into the dredged channel. In effect, when you dredge a channel, you are artificially over-steepening the intertidal profile, which may begin a natural adjustment to regain stability.

In addition to the two points above, modification of flood and ebb directions can cause modifications to sediment deposition and erosion patterns, as we have seen earlier. Under normal conditions, sediment within the estuary may be moved around within the system. Where dredging has occurred, the material is often moved outside the estuary, representing a net sediment loss to the system and becoming unavailable for redistribution.

Asides from dredging, tidal waters are often seen as safe places for sailing and power boating. Constant wash from boats causes serious damage to marshes and mudflats. This problem has been dealt with where it is serious, such as in Southampton Water, the Orwell estuary and Lymington, all in England. By imposing speed restrictions, either generally or at certain times of the tide, it is possible to reduce disturbance to a minimum.

5   *Embanking and land claim.* The intertidal areas of many estuaries have been greatly reduced owing to the practices of land claim and sea wall construction, as we have seen earlier. Unlike on wave-dominated coasts, sea walls are rarely constructed on tide-dominated coasts for purposes of defending the land against erosion by the sea directly; rather they tend to be used to prevent flooding from areas which farmers want to claim and use for agriculture. Hence, the form of sea wall found in estuaries is fundamentally different from that found on open wave coasts. Instead of large concrete structures, earth embankments are used to protect the land. With little wave

---

BOX 3.3 The relationship of windiness to marsh erosion/accretion cycles in the Severn estuary

Wind variation acts as a modifier to the wave and tidal regime of the Severn estuary. Increases in wind strength will produce increases in wave height and wave energy, as well as the possibility of producing a storm surge. The most frequent wind direction is from the west or south-west (Hannel, 1955; Walker, 1972), with secondary modes occurring from the north-east and from the south. The north-easterly trend arises as a result of the development of high-pressure systems to the north of the region, particularly in the spring, when the prevailing wind direction is predominantly north-easterly (Walker, 1972). The predominance of westerly and south-westerly winds is particularly important for the system because it is also the direction to which the estuary opens; that is, a wind from this direction will blow directly up the estuary.

The most dramatic effects of wind are when its speed approaches gale force (Force 6), or stronger. At such times, surface waters show wind-generated waves which are a prime cause of sediment erosion, in respect of subtidal sediments above wave base, and of intertidal sediments and the sea-facing cliff line of the salt marsh sequences. Such wind strengths occur for only 1.8 per cent of the time during which winds are observed (Hannel, 1955) and of this 1.8 per cent, 41 per cent occur in winter, 33 per cent in spring, 22 per cent in autumn and 4 per cent in summer. As winter is also the time when the salt marshes have least protection by vegetation cover, this tends to be the period of most large-scale erosion. Also, the autumn gales erode large blocks, weakened through deep summer desiccation. Again, the most frequent prevailing direction for gale-force winds is from the west or south-west.

In terms of erosion potential, the frequency of south-westerly winds appears to play an important part (Allen, 1990). Historical records show that there are distinct periods when south-westerly wind frequency is high, and periods when it is relatively lower (Lamb, 1982). Periods of high south-west wind frequency appear to occur in the early twentieth century, the mid- to late nineteenth century and the early eighteenth century, and so these may be expected to be times of high erosion. The sediment accumulation within the system occurs in the form of distinct erosion/accretion cycles. The main periods of salt marsh erosion occur during the early twentieth century, the late nineteenth century and the early eighteenth century, a good correlation with the westerly and south-westerly wind frequency.

In addition, other work indicates that measurements of fallen trees, blown over in gale conditions, indicate a west–east alignment. Where root structures are evident, these clearly indicate a falling direction towards the east.

activity, the energy received at the wall is lower, and thus they need not be as structurally significant.

6   *Loss of vegetation.* Increased environmental stress can cause vegetation dieback. These stresses can be either natural or anthropogenic in origin, but regardless, the effects are the same. The presence of vegetation on sediment serves two main purposes:

    (a)   The roots bind the sediment and make it less susceptible to erosion.

    (b)   The above-ground vegetation acts as a baffle, trapping sediment as water flows through it at high tide.

Thus the presence of vegetation on a marsh is important in the stability of that environment. Excessive dieback has largely been restricted to *Spartina* marshes, *Spartina* being a pioneer species which is one of the first colonisers of mudflats, but vegetation loss can also arise from increased channel incision and widening.

7   *Reduction in sediment supply.* Sediment in estuaries can be derived from two sources. It can either be brought in from the sea, or brought down the rivers. We have seen how sediment can be reduced from the sea, by means of coastal defences, and also from the removal of sediment by dredging. There is ample evidence, however, that a significant component of 'missing sediment' derives from the fact that there is less being brought down the rivers, the problem arising in a way similar to the reduction of input from the coast, namely that river engineers have undertaken significant river engineering projects and river entraining so as to minimise river erosion. This issue will be explored further in Chapter 6.

8   *Sediment extraction.* Many British marshes have been used for turf, sea bank construction, and minor sediment removal. If carried out correctly, this can be done with no damaging effects to the marsh. In some places, however, notably in the Medway estuary, south-east England, clay removal for brick- and cement-making has removed the majority of marsh sediment (Box 3.4).

9   *Problems of accretion.* We have discussed the problems of wave- and tide-dominated coasts as being a result of erosion. There are also very real management problems on tide-dominated coasts which arise out of the problem of accretion. We have already seen that fine sediment will accumulate where tidal currents slacken. There have been many examples of this occurring owing to changes in channels, etc., and it is quite possible for a once silty environment to become muddy. Once mud starts to accumulate and build up, it starts to be colonised by vegetation, and thus become a marsh. This is fine unless the area in question is used for some other purpose, notably as a beach. Southport, on the mouth of the Ribble estuary in north-west England, has had major problems with encroachment of *Spartina*. Similar problems have also occurred at Lindisfarne, in the north-east of England, and at Grange, on the banks of the Kent estuary, Morecambe Bay. From a management point of view, the development of a marsh is a great way to defend the coastline naturally. From an amenity point of view, a marsh is no substitute for a beach of golden sand.

BOX 3.4 **Removal of marsh clay in the Medway estuary for brick-making**

Typically, the loss of large areas of marsh is associated with natural agents of erosion. In the Medway, however, large quantities of marsh clay have been dug for use in brick-making, particularly during the latter part of the nineteenth and the early part of the twentieth century (Medway Ports Authority, 1975). Extraction was prohibited from the marsh edge but the central parts of the marshes were dug to considerable depth, cuts being made to channels to allow the removal of the dug clay by barge, but also allowing water to enter the workings.

Tight legislative controls were exercised over this industry with the issuing of licences by the Medway Conservancy, in order to govern the locality and quantity of clay removed. Despite these controls, however, many instances of illegal extraction did occur, resulting in the loss of considerable marsh areas. The areas in which extraction was licensed occur throughout the Medway and are detailed in the table below. Although the last licence issued was dated 1931, extraction at this time was much reduced compared to that at the turn of the century, owing to the move to alternative sources of clay inland. Despite this, it was not until the mid-1960s that the last shipment of clay was made.

The importance of this activity in relation to the loss of marsh is not clear, and natural agents of erosion in response to dredging and sea bank construction have also played a part. However, some areas have undergone significant loss of marsh by the removal of clay (estimates put this loss in some areas as being as high as 90 per cent; Kirby, 1990). The amount of clay removed is also unclear, but estimates for one company at one site (West Hoo Creek) suggests 1.4 million tons between 1881 and 1911. Over the Medway as a whole, many companies extracted from many sites, and so the total loss of marsh can only be surmised.

Licences granted for clay digging and associated activity in the Medway

Licence no.	Date	Area
	1881	Hoo Saltings (*Digging and extraction*)*
114	Aug. 1884	Colemouth Creek (*Digging*)
159	Aug. 1895	Otterham Creek (*Extraction*)
162	Apr. 1896	Medway Saltings, Upchurch (*Digging*)
164	Apr. 1896	Twinney and Millford Hope Saltings (*Digging*) and Twinney Creek (*Extraction*)
167	Jul. 1896	Halstow Creek (*Entrance to marsh*)
171	Oct. 1896	Colemouth Creek (*Entrance to allow clay extraction*)
172	Nov. 1896	Mackay's Court Saltings (*Digging*)

175	Apr. 1897	Bishop's Marsh, South Yantlet Creek (*Digging*)
182	Apr. 1898	Millford Hope Saltings (*Digging*)
189a	Dec. 1898	Bishop's Marsh (*Digging*)
191	Mar. 1899	Sharpness Saltings (*Digging*)
196	Sep. 1899	Haversham Marsh, Otterham Creek (*Digging*)
199	Dec. 1899	Barksore Manor Saltings (*Digging*)
204	Mar. 1900	Rainham Creek Saltings (*Digging*)
229	Sep. 1903	Bedlam's Bottom, Stangate Creek (*Digging*)
230	Sep. 1903	Colemouth Creek (*Digging*)
234	Jul. 1904	Funton Creek (*Digging*)
237	Mar. 1905	Chitney Marsh, Stangate Creek (*Digging*)
238	Mar. 1905	Stoke Creek (*Digging*)
240	Feb. 1906	Slede Creek (*Digging*)
264	Dec. 1920	Gillingham Saltings (*Digging*)
266	Apr. 1921	Stoke Saltings (*Digging*)
269	Mar. 1922	Slayhill Marshes (*Digging*)
296	Jun. 1929	Oakham Ness (*Digging*)
313	1931	Bayford Marsh (*Digging*)

Source: Data from Medway Conservancy (n.d.), 'Terrier of Rents'. Unpublished document, Sheerness Port Authority, Sheerness, Kent except: *Information from MacDougall, P. (1980) 'The Story of the Hoo Peninsula'.

◆   ◆   ◆

As we have seen, there are many issues associated with management on tide-dominated coasts. Not all of the issues raised will occur in every situation; the relative importance of each management problem will vary according to region. Storminess, for example, appears to have been important in the estuaries of western England, while embankments and land reclamation have been significant in estuaries in many countries. There is a broad range of environmentally specific problems, which all require a sensitive approach. The need for sensitivity is doubly important because these areas, especially estuaries, are important wildlife sites, not least because of the fact that mudflats and marshes are estimated to be among the most productive habitats in the world (McClusky, 1989).

Any management technique has to recognise the importance of an area for wildlife. The Wash, in eastern England, for example, is an internationally important site for overwintering bird populations, as are Morecambe Bay and the Severn, the latter being home to the famous Slimbridge wildfowl reserve on the west coast. Similar importance can be assigned to other areas of marsh, such as the Bay of Fundy, Canada, and Chesapeake Bay in the USA, although any estuary will have some wildlife importance. For this reason, it is necessary when dealing with defence issues to take a long-term and holistic overview of the whole environment. Management techniques used on tide-dominated coasts can take various forms, as will now be discussed.

Offshore breakwaters

We have already seen the use of offshore breakwaters in the context of wave-dominated coasts, and how these structures can be used to aid the build-up of coastal sediment to reduce wave attack on the beach. It is possible to use the same methodology in the context of tide-dominated coasts. In Essex, on the east coast of England, there have been ongoing problems of marsh erosion for many years. At the northern end of the Dengie peninsula, the offshore breakwater consists of a series of old Thames barges, which act as a wave baffle and reduce tidal scouring on the ebb tide. The result has been that there is an area of mudflats which has accreted rapidly, and will, therefore, provide an area where water depth will be lower, and so wave and tidal scour will be reduced at the sea wall.

Barrages and barriers

Barriers and barrages are designed to protect low-lying coasts from flooding, often following the extensive reclamation of an area which has subsequently undergone relative subsidence due to dewatering and the continued accretion of sediment outside the walls. Barrages may be of several types. Those designed to serve as flood protection, such as the Thames Barrier, are in place only when flooding is predicted. Others are permanent structures which also serve to generate power, and keep salt water out of an area. The concept of barrages overlaps with another key theme, that of power generation, and the environmental problems of these structures are discussed fully in Chapter 4.

Sediment recharge

As with sandy beaches, mudflats can also be recharged with sediment. In this case, however, because mudflats are largely on wide foreshores which are gently sloping and involve cohesive sediment, different criteria have to be met. The recharge of sandy beaches needs to employ the correct grade of sediment because waves move sediment around in the short term to compensate for wave activity. For this reason, the sediment has to be sufficiently mobile. Mudflats, because they are cohesive, respond at different rates, and so it is not as critical for the stability of the coast to maintain the correct grain size distribution. Instead, the prime motive for recharge is to increase the defence potential, and this may be done by either slightly increasing the grain size to make sure it stays on the flat, or increasing the cohesiveness of the mudflat sediment. At Hamford Water, Essex, eastern England, slightly coarser sediment than that already present in the environment was used to recharge two sites at Foulton Hall and Stone Point (Pethick and Burd, 1993). Material was derived from dredging operations from a nearby harbour, therefore successfully reusing surplus sediment and keeping it within the near-shore zone.

Because of the contrasting nature of muddy and sandy sediment, recharging mudflats is more complex than for beaches. Muddy sediments do not develop an equilibrium immediately, but need time to dewater and consolidate, and develop an infauna of micro-organisms. It is only after such processes that the full cohesive strength is realised. This means that mudflat recharge is not a rapid process, because it is necessary to protect the recharge sediment from waves and tides while they are settling and dewatering.

Basically, a series of deposition ponds needs to be constructed in which sediments can settle. When this has happened, water is pumped out, and the sediment left to dewater and consolidate, with increasing amounts of tidal inundation allowed.

ENVIRONMENTAL IMPACTS

Because mudflats are complex ecosystems, their organisms are sensitive to large and sudden inputs of sediment. A covering of 1 or 2 cm can be coped with by upward migration; half a metre or so would swamp the infauna and lead to mortality. This would have serious knock-on implications for bird and fish life and also for sediment stability because secretions from infauna also help in making muddy sediment more cohesive. This has to be allowed for in the calculation of the area to be recharged. In areas where birds are important, for example, recharge will have to be added over a period of time to allow the infauna to maintain its position in the sediment profile.

The materials used also need to be tested for toxicity. Often, the technique employs locally dredged sediment, which can be highly toxic (see Chapter 4) if it comes from ports and harbours, especially if shipbuilding or petrochemical processing occurs at the source. It is, however, a good way of using dredged sediments, as in the case of Hamford Water, Essex. The sediment on the recharged flat remains within the sediment budget, and so the use of local dredgings means that this material is not lost to the system.

Increasing natural sedimentation

A second way to build up mudflats is to allow sedimentation to be a natural ongoing process, rather than pump sediment from another source. This is particularly the case where there is a plentiful sediment supply, or where parts of the ecosystem, such as infauna, are especially prone to damage. Allowing the gradual build-up of material enables infauna to adjust and compensate for sediment deposition, thus removing the danger of inundating the fauna during the pumping process.

The rate of deposition is largely controlled by the velocity of the currents and vegetation cover, and therefore to increase deposition, it is necessary to decrease these current velocities which keep the sediment in suspension. This may be achieved in one of two ways. The first is to construct polders of brushwood groynes (Plate 3.3), as in the case of the Schleswig-Holstein method. Water can flow into the area, but velocities are reduced by the ponding effect, thus allowing sediment to settle out. The fact that

PLATE 3.3   The use of brushwood groynes to increase sedimentation in the intertidal zone (Wallasea Island, Essex, UK)

the materials used for the construction allow water to move through, means that water can slowly seep out, without generating significant tidal currents. Accretion can be aided by trenching the mudflats to increase frictional drag on the water, by increasing the roughness. This method of foreshore accretion is widely used, having been developed from experiences on the Dutch coast and subsequently used in other countries. A second way of increasing natural sedimentation is to reduce current velocity by increasing the baffle effect caused by vegetation. *Spartina* sp. spreads rapidly to colonise mudflat areas and, because of this, has been commonly planted as a means of trapping sediment and increasing the elevation of such mudflat areas.

ENVIRONMENTAL IMPACTS

In the Schleswig-Holstein scheme, sedimentation is largely dependent on natural supplies of sediment. However, the process is still artificially increasing the rate of sedimentation, and so it is still possible to swamp infauna if sedimentation rate increases too much. Similarly, the amount of dewatering which occurs is also important. The entrapment of sediment can lead to a sediment with high water content, which proves poor for habitat formation. There is consequently a danger of developing an area with contrasting sediment properties when compared to un-recharged sites, each having its own suitability for infauna, and its own stability to resuspension and erosion.

The planting of *Spartina* can also lead to environmental impacts. Because of its invasive nature, it can swamp important feeding grounds, which will have secondary effects on wildfowl. Also, because it is so dominant, it can alter local ecology by ousting native colonising species.

Flood embankments

As with all coastlines, it is quite possible to defend against flooding by constructing defences in the high intertidal area. Wave activity is not as much of a problem on tide-dominated coasts as on wave-dominated coasts, and so the scale of defence construction need not be as large. Typically, earth embankments are used, often constructed by taking sediment from the fronting salt marsh as discussed in the section on the environmental implications of land claim (p. 58).

ENVIRONMENTAL IMPACTS

Flood embankments, because they are designed to prevent any interaction between land and estuary, by definition form a barrier between the two environments. Under conditions of sea level rise, habitats would naturally migrate landwards in order to maintain their position relative to the tidal frame. Where defences are present, this cannot happen, and problems of coastal squeeze result.

The construction of the defences can also have a long-term impact on the estuary. Both the methods used in obtaining the mud to build the walls necessitate removal from the marsh. Scraping from the whole surface lowers the level by centimetres, and vegetation can normally re-establish itself fairly quickly, although possibly reverting to a mid marsh from high, or low from mid. Similarly, the digging of borrow pits is normally only a temporary feature, although they can take around fifteen years to infill (Osborn and French, in prep.). Successful infilling, however, does not always occur, owing to poor management of the pits' construction. Digging pits too deep can mean that their bases are lower in the tidal frame than the lowest marsh communities, hence the re-establishment of vegetation will not occur. In addition, over-deepened pits also act as stores of water, which, on ebb tides, discharge large volumes of water through the creek network, sometimes leading to erosional problems within the creeks (Plates 3.1a and 3.1b).

In wave-dominated environments, wave reflection is important in relation to beach scour and loss of sediment. In tide-dominated environments, however, this phenomenon is less common, first because wave action is much less, and second because the sediment is more cohesive and often vegetated, thus making it more resistant to erosion.

Managed realignment

There is growing evidence to suggest that instead of always trying to keep the coast-line static – that is, hold the existing line of defence – it may be better to allow the sea to erode parts of the coast – to abandon land to the sea. This method is referred to as managed realignment (also called managed retreat or set-back). The idea of allowing parts of the coast to erode, or revert to marine habitat, does perhaps run contrary to the term 'defence'. However, 'defence' relates to successfully protecting the hinterland, and so managed realignment, as a method of protecting the hinterland from erosion, is important. We have seen that for the best defence, it is necessary to operate with nature as much as possible, and if nature is saying that a coast should move inland, then working along those lines will lead to better longer-term stability than constantly battling against the natural forces of tides and waves. The concept of managed realignment combines this rationale behind the need for coastal defence, the need to replace habitats lost as a result of coastal erosion, and also the need to compensate for problems of sea level rise and coastal squeeze. This process is particu-larly important in areas which experience coastal squeeze that has resulted in the reduction of intertidal width and loss of habitat and wave attenuation ability. This reduction in width reduces the buffering protection afforded by the intertidal area fronting the sea wall. Increasingly, the sea defences become destabilised as the toe of the wall is exposed and eventually becomes undermined.

We have seen that throughout history, humans have changed the outline of the coast by large-scale reclamation, to gain areas of flat, cheap land for agriculture and development. Throughout history, this process has accounted for an estimated 25 per cent of intertidal land in estuaries, resulting in the destruction of large areas of the coastal zone, world-wide. The continued exploitation of the coast has important implications for habitats and wildlife, as the more land claim and defence works which go on, the more tightly squeezed the intertidal zone becomes. As sea levels continue to rise, so this squeeze will become greater as intertidal habitats are pushed further up against hard sea defences, such as sea walls, and create increasing instability in the coastal zone. The accommodation of this natural process of sea level rise can be met in one of two ways:

1   Natural geomorphological processes are fundamental to coasts being able to adapt to change; thus natural features such as a beach or shingle bar can move and adapt to new wave and sea level conditions. Habitats can change and adapt alongside such changes, provided that freedom of landward movement is maintained.
2   These new habitats can be artificially created by allowing the movement of the intertidal zone landwards by deliberately removing the line of hard defence. This process is known as managed realignment.

The general concept of this technique involves the creation of new habitat to act as a buffer in the reduction of wave energy. We have already seen how the fore-shore can be recharged and built up in front of defences by advancing the foreshore forwards into the sea. Managed realignment involves the landward movement of a sea defence structure and the promotion of new habitat creation in front of this new line of defence. The land between the old and new defences then forms a new part of the intertidal zone which is more able to respond to coastal processes, and thus reduce the effects of coastal squeeze. Goodwin and Williams (1992) have reviewed this technique, combining habitat creation through both coastline advance and coastline retreat, for example along the Californian coast.

Up to now, this technique has been used in the 'artificial' creation of salt marsh. In the UK, there are several sites where managed realignment has occurred. One of the most heavily studied areas is the Essex coast, where this practice has been used at four locations (Old Hall Marsh, Tollesbury; Northey Island; Fambridge; and Wallasea Island; Boorman and Hazelden, 1995; Turner, 1995). However, it can also be shown that, historically, evidence of naturally occurring realignment has happened through storm breaches which have not been repaired. In the UK, a period of increased storminess around the turn of the century meant that there were many cases of sea defences being breached around this time. Not all these were repaired, particularly where the inundated area had been previously reclaimed and the flood inundated land to the previous sea defence. These sites provide evidence for how this technique, when artificially managed, can be of value in creating new habitat. Some of these sites have been more successful than others. Pagham Harbour, Sussex, was fully reclaimed at the end of the nineteenth century, but breached at the onset of the twentieth. The inundation of the sea initiated new marsh growth and an active intertidal zone. This area is still accreting and developing marsh growth, and thus has proved to be a successful retreat site (French, 1991).

At Fambridge, on the river Crouch in Essex, a similar breach occurred at around the same time. This flooded land back to a previous defence, and provided healthy marsh protection for many years. At the present time the marsh is eroding, but it can be argued that the set-back area successfully protected the coast along that stretch of river for eighty years, and is still providing some degree of protection today, while at other sites in the river, the marshes are narrow and close to the sea wall.

It can be seen from these natural set-backs that this principle can be used to provide low-cost flood defence, or provide extensive areas of habitat creation. Increased flood defence is provided by the presence of an intertidal profile which dissipates wave energy.

ENVIRONMENTAL IMPACTS

The most obvious environmental impact is that areas of dry land revert to the intertidal zone. The loss of terrestrial habitats is compensated for by the addition of intertidal ones, and it can be argued that terrestrial habitats have the greater freedom to migrate laterally than do intertidal ones.

It is clear that some environmental conditions favour realignment, while others do not, so that this method of coastal defence can be considered along the same lines as others, as far as environmental suitability is concerned. The most successful schemes have occurred where the following four parameters were available:

1   The elevations of the original site were suitable for the desired habitat – i.e., higher for marshes, lower for mudflats.
2   The site was adjacent to an existing source of flora and fauna – i.e., next to areas of marsh or dune vegetation to allow migration.
3   The hydrological conditions of the site were maintained to prevent scouring and allow sufficient water stand to allow sedimentation.
4   The sites had been well researched, and advice had been obtained from suitably qualified experts.

We have already seen that when defence measures are being considered, each option is subject to a cost–benefit analysis. Managed realignment options also need to be judged by the same criteria. There is expense involved in relevant engineering works, especially if new inland defences are required. Similarly, the creation of new habitat is also an expensive undertaking, if this has to be artificially encouraged or enhanced. Offset against these costs, however, is the value of the new habitats, both as wildfowl refuges and as coast protection measures. In the USA, estimates indicate that coastal wetlands are worth around £18,500–£26,000 per hectare, based on their value in coastal defence provision. Balanced against that, the same habitat costs between £3,200 and £44,500 per hectare to create artificially, suggesting that the economics are better favoured in some areas than others.

Finally, as with any defence, it is important to carry out all actions with consideration of other parts of the coast, and the relevant sediment budgets. As most of these schemes are based on estuaries, this means that it is important to bear in mind the impact that these actions will have on estuarine processes. The amount of sediment available to an estuary will remain the same whether a realignment scheme goes ahead or not. Thus increased deposition in such an area will mean that sediment will not be deposited in other areas. This could trigger off new erosion, or modifications to channel morphology.

The issues raised by managed realignment are significant. This approach is becoming an increasingly common method of coastal defence, and we will be returning to the process later.

Large-scale hydraulic remodelling in estuaries

Many of the problems which cause erosion on tide-dominated coasts are the result of currents. In addition, many of the management problems which arise are deep-seated, and local-scale measures are not likely to solve the problem overall. Major problems such as sea level rise, tidal asymmetry, and negative sediment budgets cannot be tackled by piecemeal measures, but could be, theoretically, by major works designed to modify the hydraulic regime of estuaries.

Measures which could be taken include the construction of barriers, weirs, submerged shoals, breakwaters, artificial headlands, and islands, all measures which we have discussed already as individual means to manage coastal systems, but which in combination could be used to modify an estuarine environment completely. Simpler measures could involve opening up new channels through marshes to alter flow or ebb dominance, or increase bed roughness. To date, however, this form of management is theory – and has not actually been put into practice.

Coastal defence management and planning

The planning of coastal defence measures

The initial trigger for the onset of defence planning will lie in the realisation of there being a problem: a large cliff fall, a strong storm, the loss or flooding of property. There are established procedures and planning stages laid down which can be followed, forming a framework within which projects are planned, investigated, and implemented in the best way to meet the demands of both the local population and the natural environment. It is really the balance between these two factors (natural environment and local population) which will decide whether the defences will be built or not.

There are six steps to the framework.

1   *Preliminary thinking*. During this stage, careful thinking is undertaken to define clearly the extent of the problem, the desired final outcome, and a list of the various ways by which that outcome can be achieved. Such aspects as minimising the risk to life will be placed at the top of the list, but the list will also include other factors such as the maintenance or enhancement of the built and natural environment – including physical, cultural, and biological aspects.

Bearing in mind the relative importance of these factors, one can then investigate the methods available. At this stage, it is also necessary to consider what will happen if nothing is done; only by doing this is it possible to judge the merits of the other options. Also, it is important to consider the effects of the proposed work on the rest of the coastline; that is, the coastal cell approach (see Figure 3.6).

Where defences already exist, the thought process will consider the merits of reinforcement or extending these. In effect, there are several categories of approach utilising the methods discussed earlier which are available at this stage:

(a)   do nothing (i.e. leave the coast alone to find its own equilibrium state);
(b)   undertake managed realignment (i.e. move the defence inland to allow the creation of new habitats);
(c)   opt for partial set-back (allowing parts of the coast to erode while protecting others);.
(d)   maintain existing defences (reinforce and repair);
(e)   build new defences (protecting previously unprotected parts of the coast);
(f)   advance seaward (i.e. offshore breakwaters).

Finally, it is important, from this early stage, to include all interested parties in relevant discussion. These could include:

- statutory and voluntary groups;
- local people;
- those with specific site knowledge;
- coastal experts;
- nature conservation bodies.

2   *Developing and appraising the options.* All options should be developed and appraised through careful, integrated planning. The options will fall within four categories:

(a)   Do nothing. Basically, take no action to maintain existing defences, improve on defences, or construct any form of defence. In other words, allow the coastline to continue to erode naturally, as in the example of Holderness. It is only by doing this type of survey that all other options can be assessed.

(b)   Manage potential risk. Rather than defend the coast, improve the warning systems to allow people time to evacuate. This option is particularly relevant to areas of coastal flooding, rather than erosion, but is important where aesthetics are important, and the emplacement of large concrete defences would detract from a tourist area, from an area of natural beauty, or where the supply of sediment along a particular coast is critically important to other parts of the coastal cell.

(c)   Maintain the existing line. Repair the current defences to maintain the current standard and line of defence. Such actions will include the rebuilding of walls, or the replenishment of sediment on beaches. This would then involve a further investigation of hard *vis-à-vis* soft engineering.

(d)   Change the existing method of defence. This may be achieved by:

   (i)   reconstructing or reinstating the existing measures with increased strength and resistance to wave attack (upgrade);
   (ii)   implementation of new measures to improve the performance of existing defences; that is, supplementing defences with groynes added to sea walls or secondary walls, etc;
   (iii) constructing a new defence seaward or landward of an existing defence, such as offshore breakwaters or managed realignment;
   (iv)   construction of new defences in areas where they were previously absent.

In addition, consultation with interested parties should continue, and the area should also be investigated for any environmental designation; for example, is it a nature reserve or an SSSI (site of special scientific interest)?

A third stage is to investigate the environmental impacts of the options. This investigation should include:

(a)   looking at the information from consultations;
(b)   assessing impacts within the rest of the coastal cell on both natural and artificial environments; that is, effects on sediment erosion, sediment transport, and deposition patterns;
(c)   identifying areas of special interest; that is, areas which need to be protected, or which contain geological or ecological sites;

(d)   consideration of impacts during and after scheme implementation;

(e)   identifying ways in which (d) can be prevented or minimised;

(f)   considering opportunities for environmental enhancement.

After carrying out these stages, it is possible to eliminate those options which will give too great an impact, leaving those which are technically sound and feasible.

3   *Choosing the right option.* At this stage, the environmental assessment needs to be formalised where the proposed works will require planning permission. The need for this basically depends on the magnitude of the project, and whether it falls within Schedule 1 projects, in which case an environmental impact assessment (EIA) is mandatory, or Schedule 2, where an EIA may be required. The EIA will produce an impact statement which assesses the impacts of the development on humans, fauna and flora, soil, water, landscape, and the interaction between these.

At this stage, all surviving options are environmentally, economically, and technically feasible. These options can be ranked into two groups with one relating to cost–benefit and the other to environmental impact. If both lists favour the same preferred option, then the decision is clear-cut. If the preferred option is different, then the decision is a matter of judgement. After the preferred option has been selected, it is now taken forward into the design stage. If, as is possible, the preferred option is to do nothing, then the process will stop here and the coast be left to erode naturally.

4   *Design.* The design stage should consider all the environmental implications of the project, and all the environmental issues raised under the consultation phase. It should also study such factors as wave reflections and effects on adjacent coasts, sediment dynamics, and what will happen at the edges of the development – the fringing effects.

The protection, conservation, and enhancement of the coast should be integral at all design stages. There should be detailed consideration of:

(a)   maintenance/improvement/creation of habitats. No project should reduce habitat diversity or leave an area ecologically impoverished, but could be made more acceptable if it includes habitat creation;

(b)   protection for important species;

(c)   retention of landscape, morphology, geological features. This is especially important when considering sea walls, as concrete is very effective in obliterating rock outcrops and coastal morphology;

(d)   protection of archaeology and built and cultural heritage, which may require hard defences, the construction of which may affect the design of the rest of the scheme;

(e)   sources of materials: where the material will come from – sand for beach feeding, rock for offshore breakwaters, etc;

(f)   retention of access to the shore by pedestrians or boats – the coast must not become inaccessible to users;

(g)   methods of construction, sources, and disposal of materials; that is, there should be no contamination of the local environment.

The development of any defence structure, as we have already seen, will include many opportunities for the environment. Schemes may provide opportunities for enhancement of the terrestrial environment, or restore habitats such as lagoons or marshes, or create artificial reefs. The development should be sympathetic with the environment, and the damage done to existing habitats always borne in mind.

5   *Operation.* The operational/constructional stage needs to be planned and carried out with particular care. There is tremendous potential for damage to the environment, and the destruction of sensitive areas. Consideration needs to be given to:

(a)   suitable timetabling to take account of such factors as flowering times, breeding seasons, etc. Floral communities need to set seed, and so damage at flowering times can reduce the vegetation in successive seasons. Also, birds need quiet and safety during nesting periods and when feeding young;

(b)   close definition of working areas to avoid compaction of beaches, and the minimising of effects on marshes and dunes. Construction should avoid sensitive areas and also restore any damaged sites;

(c)   effective communication and understanding to lead to proper assessment of construction impacts.

During construction, it is easy to cause accidental damage to parts of the shore-line, and it is important to have a full understanding of the implications of these events. The upper shoreline, for example, is an important habitat for many bird and plant species, and so disruption must be kept to a minimum. Similarly, intertidal areas are important for marine invertebrates, including fisheries.

6   *Post-project appraisal.* After completion, the appraisal of a project is critical to allow the amelioration of unforeseen environmental problems, and to improve the design and implementation of future schemes. This is important because although the reaction of the shoreline to defence works can be modelled, subtle changes in wave and wind activity can alter the criteria on which these models are based. An example of this is given by Fox (1997): wind directions caused problems in the Sidmouth offshore breakwater scheme, resulting in the offshore transport of sediment. It is important to remember throughout that the predicted impacts are exactly that – predicted. It is also important to remember that knowledge of the coast is not spatially uniform. Some areas are well studied, others less so, and so the information base which forms the basis of predictions is also variable in quality and completeness.

Post-construction monitoring is essential for all defence works, looking at how efficient the works are in solving the problem which they set out to solve, and also the effects further along the coastline. An environmental and engineering audit needs to be carried out after the implementation of the scheme, based on the comparison between baseline surveys and what is actually observed.

Mitigation action should be put in place if required if it becomes evident that the environmental objectives of the scheme are not working. This can take many forms, but may need to respond to loss of beach sediment, erosion at another locality, or, most severely, the failure of the defences themselves.

Responsibility

The procedure described above is all well and good, but what needs to be determined is exactly who has the responsibility for initiating such schemes. Clearly, in such situations we are verging on the grey area between coastal engineering and coastal zone management. Increasingly, in many countries, the latter is gaining in importance, and is an integral part of the whole coastal planning process. In the UK the system has not developed as much, and the system of responsibility is still fairly complex.

In the UK, ultimately it is the Ministry of Agriculture, Fisheries and Food (MAFF) that is responsible for providing grant aid to various authorities for capital projects. These authorities vary, largely according to the role and function of the defences. Flood or sea defence projects – that is, those necessary to protect property from being flooded by the sea – are the responsibility of the Environment Agency (EA) formerly the National Rivers Authority (NRA), under the Water Resources Act 1991. These flood defence functions are controlled by ten flood defence committees, chaired by a ministerial appointment and made up of local council representatives. It is also possible for groups to act independently. Under the Land Drainage Act 1991, local landowners (including Railtrack, the company that manages the UK's railway lines and stations) may also undertake such works, and so in this respect different groups can serve the same defence function without any consultation.

Coast protection – the prevention of erosion – falls within the bounds of maritime district councils. Under the Coast Protection Act 1949, these councils are empowered to carry out such works as are necessary to protect their administrative area. Because these works have significant implications for the behaviour of the coastal cell, maritime councils are required to consult with other groups, notably the NRA, neighbouring councils, the relevant county council, the harbour authority, conservancy and navigation authorities, and fisheries committees; that is, all those groups upon which defence activities can have an effect. Any scheme, excluding emergency repairs, needs MAFF approval anyway, and before this approval can be given, other bodies, such as the Crown Estate Commissioners, the Countryside Commission, English Nature, the Countryside Council for Wales, all need to be informed, and all schemes have to be notified to the general public. Similarly, other landowners can also undertake protection works, subject to notification of the relevant council, and the conditions outlined above.

The co-ordination of works is the responsibility of the local authorities, although the Environment Agency is required to exercise general supervision over all matters relating to coastal defence. Frequently, however, much of the co-ordination is already worked out owing to the establishment of regional coastal groups – which have produced framework documents detailing defence strategies for many parts of the coast.

Despite the responsibility of local authorities for the flood and coastal defences, these powers are permissive, rather than mandatory. This means that the authorities are not obliged to prevent flooding and erosion, and have the option of allowing natural processes to operate; that is, to do nothing.

MAFF will provide funding to schemes proven to be environmentally acceptable, technically sound, and economically justifiable. Funding is available both for sea·wall construction and for recharge schemes. For schemes outside these controls such as residual costs, or non-grant-aided schemes, finance is from local funds such as regional environmental agencies (obtained from levies of local councils), or from local authorities themselves.

Governing legislation

We have seen that the EA and local authorities are empowered to provide coastal and flood defences. This does not mean that they have a free hand to do what they like. The 1991 Water Resources Act (section 16) obliges the EA and ministers, and the 1991 Land Drainage Act (section 12) obliges the EA, internal drainage boards, and ministers, to:

◆ further the conservation and enhancement of natural beauty consistent with any enactments relating to their functions;
◆ further the conservation of wildlife and geological/morphological features of special interest, consistent with any enactments relating to their functions;
◆ have regard to the desirability of:

(a) protecting and conserving buildings, sites and objects of archaeological, architectural, or historic interest, and
(b) preserving public rights of way and access to the foreshore;

◆ take into account the effects of any proposals on the preservation of rights of access, on wildlife, features, buildings, sites, or objects of interest.

Basically, while the authorities concerned are not forced by law to protect areas from erosion and flooding, if they choose to do so then they are governed by strict controls and guidelines as to what they have to do and what they need to take into account.

◆ ◆ ◆

We are now in a situation where we have studied several aspects of coastal defences:

1 the sorts of problems which occur on the coast, and the environments in which they occur;
2 the methods which can be used to prevent erosion and flooding, and the impacts which these can cause;
3 the legal and responsibility framework in place to oversee these works.

It is not ideal to have the whole coast shrouded in defences, and so some selectivity has to occur. On the basis of cost–benefit analysis, if the cost of defending the coast is greater than the value of the land which is being defended, then no defences will be built. Isolated farms tend to be regarded as developments which are not worth as much as the cost of the sea defences, and so are not defended. Major infrastructure or industry is of critical national importance, and is of value clearly in excess of the cost of sea defence construction and so is more likely to be defended.

Managed realignment: a universal solution in soft engineering?

As yet, there is one main aspect which we have not considered in detail, although the concept was introduced when we discussed the ways in which tide-dominated coasts may be defended. This approach to coastal defence is one of the most important contemporary issues in coastal management. Letting parts of the coastline erode, either naturally (do nothing), or in an encouraged and controlled way (managed realignment), is perhaps the best way of protecting the land. Increasingly, this approach is favoured, and represents a new concept in human impact on the coast. This impact is different from those which we have already considered because instead of being negative, the process of realignment actually creates habitat and manages the coast in a way sympathetic to the natural processes. For this reason, the option is often viewed as the ultimate saviour in coastal defence, and one that can solve all the problems of traditional methods. While this may be true to some extent, the technique has its advantages and disadvantages along with any other technique.

Both the concept of doing nothing and the realignment of defences incite great passion and feeling – especially among people who are directly affected. The scientific reasoning behind it is sound: by defending the coastline, we may set up sediment deficits which promote erosion elsewhere, and so make the problem considerably worse than it originally was, but by encouraging an increase in intertidal area, or by allowing the artificial input of sediment, the coast will develop its natural line of defence, and so, in effect, manage itself. The complication lies in the fact that coasts have been very popular for human settlement and industrial growth, and so in many cases it is not possible to allow them to erode. The background to these theories has already been explained. Far from being such a universal solution, however, managed realignment is not the answer to all coastal defence problems. There are important site requirements which need to be met before such schemes can go ahead, and so not all areas of the coast are suitable.

The managed realignment option: suitable locations

As with any defence, it is important to carry out all actions with consideration of other parts of the coast, and the relevant sediment budgets. As most of these schemes are based on estuaries, this means that it is important to bear in mind the impact that realignments will have on estuarine processes.

Tidal prism

The most critical impact will be on the tidal prism. In some cases, the opening up of an area of sea wall, and the subsequent inundation of land, will allow more water to enter and leave the estuary during the tidal cycle, thus increasing the tidal prism. This could result in increased erosion as current velocities are increased. Conversely, water levels may become lower owing to the flooding of greater intertidal areas.

Estuarine morphology

The ultimate aim is to free up the intertidal zone to respond naturally to coastal processes, with the long-term aim of returning the estuary to a more natural shape. As, typically, estuaries are naturally funnel-shaped (wider at the mouth and tapering inland), realignments in the lower estuary should go back further than those in the upper estuary and hence greater land areas are required. The best way of achieving this is to undertake one massive retreat, covering the whole of the estuary, although this is likely to be impractical in most cases. Consequently, a series of smaller schemes is often carried out, as part of an overall strategy.

Site history

The most suitable sites for managed retreat are those which were once salt marsh before being reclaimed. This is because the soil chemistry and structure are likely to be more suitable, and also because if marsh had once grown on a site, then it is clear that this site is suitable for marsh growth. If, conversely, the site had never supported marsh growth, then there must be some reason for this, and it might be better to avoid such sites.

Surface elevation

The elevation of the marsh surface is the single most important factor contributing to long-term success in marsh creation. I have already stated, with reference to the Severn estuary, that in being claimed, the claimed marsh level falls relative to the unclaimed marsh owing to dewatering and continuous accretion to seaward. This means that the longer the marsh has been claimed, the greater the height difference is likely to be, and the lower the marsh community which is likely to develop. In some cases, the level may be lower than that at which pioneer vegetation can become established, meaning that only mudflats will develop. The higher the marsh elevation is, the fewer the times that it will be covered by the tide, meaning that:

1   more mature marsh vegetation can colonise;
2   there will be fewer tides to erode sediment from the surface, and lower wave energy;

3   there will be longer time periods between tides to allow compaction of the sediment;

4   greater settling means greater dewatering, and subsequent increases in the success rate of seed germination.

The best method of determining whether the claimed surface is at a suitable level is by relating it to those elevations already present on the existing marsh. Lower elevations are more likely to revert to mudflat, until accretion is sufficient to allow marsh plants to colonise. This may be desirable in some cases. If not, the level can be artificially raised by sediment pumping, to increase the elevation sufficiently to allow the more immediate colonisation of flora.

Surface gradient

Marshes will develop on flat land, although a gradient will allow a more diverse marsh community, in that different levels of tidal inundation will encourage a range of low, mid, and high marsh species. Many of the breaches which occurred naturally around the turn of the century tend to be over flat land because they have been farmed. The resulting marsh is consequently typified by a low species diversity, resulting in a rather low-grade marsh.

A sloping site, preferably grading into terrestrial vegetation, is ideal. The slope encourages a diverse flora, and reduces the management needed. The merging with freshwater vegetation allows the development of transitional communities. Such sites, however, are rare, especially because managed retreat is most needed in sites which have undergone extensive reclamation, and which therefore are often those with several lines of defence. In these circumstances land is likely to be set back to a previous line of defence, rather than to high ground.

In reality, increased habitat diversity and, therefore, greater biological diversity and conservation importance are generally achieved if:

1   the landward limit of the site grades naturally into terrestrial habitats, allowing the creation of rare transitional plant communities;

2   a natural slope is present allowing a range of marsh communities to colonise; or

3   the surface is artificially graded to produce a range of elevations, to imitate the effects of natural marsh development.

Research in the USA has shown that slope is important in determining the types of vegetation which are established on slopes. The work found that slopes of between 0 and 2 per cent were generally to be recommended, while later work indicated that the optimal results are obtained on slopes of between 6 and 7 per cent. Sites in the UK which have undergone retreat following natural breaches have typically been in the region of 0.1 per cent.

The importance of slope has been shown on the experimental managed-retreat site at Northey Island in the estuary of the river Blackwater in Essex, south-east

England. This experiment was backed by sloping land, with the result that a range of new plant communities have become established, with increasing diversity towards the back of the site, and higher elevations.

Sediment characteristics

The sediment of the new habitat will govern the success with which plants colonise. It is important that the deposited sediment is of the right grade to encourage plant colonisation. Provided that there is sufficient suspended sediment to allow accretion, this will provide the basis for future plant establishment. As long as the rate of accretion is not too rapid, so as to smother the plant communities, the continued deposition of sediment will allow the continued building up of the marsh surface. Developing root systems will then bind the sediment to form a more resistant sediment surface.

Creek networks

In natural systems, creeks establish as an important part of the inundation and drainage of the marsh. They play an important part in distributing sediment over the marsh surface, and so when establishing a new marsh, it is beneficial to have some form of creek system. In claimed land, it may be possible to reopen the original system, either by allowing natural processes to do it, or by excavation.

If no system is present, again natural erosion may cut one, although this is a long-term process and may hinder the establishment of vegetation. Alternatively, an artificial drainage network can be.cut, based on equilibrium geometry, which is itself dictated by tidal characteristics, tidal prism, sediments, accretion rates, and vegetation types. Neither of these options is ideal, but may be the best solution. The best way is to allow the site to develop its own system.

Tidal hydraulics

When any realignment project is being undertaken, it is important to know what size of hole should be created in the old defence to allow the sea to invade. The best method may be to remove the whole of the wall, but where this is not possible, it is important to have a breach of sufficient size to allow the site to achieve an equilibrium state. The main controlling factor is the amount of water moved into and out of the site on each tidal cycle, the tidal prism. If the breach is too small, the velocities at the breach will cause scour and erosion, and the net loss of sediment. If it is too large, then the currents will be slack, and the breach could become clogged with sediment.

The location of the breach is also important. Work in the USA has demonstrated that the most important elements to consider are:

1   The breach should be located at the natural mouth of a creek; therefore, flow in an existing channel will move material onto and off the marsh.
2   As the new creek deepens at the breach, the potential loss of habitat outside the mouth through ebb tide scour should be taken into account.
3   The breach should have sufficient shelter from wind and waves.
4   One wide breach is more suitable than many small ones, because if two or more are present, flow can enter through one and leave via another, thus failing to transport sediment onto the marsh. Multiple breaches may be suitable if the creek system of the marsh is established into a series of watersheds.

Managed realignment is becoming more popular and is increasingly being regarded as a suitable way of protecting tide-dominated coasts in a way which maintains a natural coastline. Brooke (1992) reports on several examples in both the USA and UK which utilise the ideas on managed realignment in conjunction with other techniques already discussed. In North Carolina, dredged material has been used to increase land levels prior to breaching and planting salt marsh vegetation. In California, many sites have been breached and used for marsh generation, with techniques involving the pumping of dredged material and natural sedimentation to increase surface height, followed by either planting or natural colonisation. In Louisiana, marsh has been created by natural processes, while in Texas, Georgia, and New York, dredged material has been pumped and planted with marsh species.

In the UK, marshes have been restored by realignment of defences along the Hampshire, Gloucestershire (Severn estuary), and Lancashire coasts. In Essex, many schemes, as already discussed, have been adopted using both pumped material and natural sediment accumulation.

Managed realignment represents the positive impact of human activity at the coast by controlled landward movement of the high water line. In other cases, this retreat can be allowed to occur naturally; that is, without any artificial management or control. In situations where land is eroding but providing an important sediment supply, leaving alone may be the best form of management. In other words, the policy of non-interference or coastal sacrifice is adopted.

The argument for this policy lies in the fact that:

1   Some coasts will cost more to defend than the land is worth.
2   Some coasts are so important for sediment supply that to protect them will cause adverse effects on the process cell.
3   Protection of the coast will mean that important habitats are lost.

At the present time, do nothing is very much an important option, but is not really well managed. People whose homes are vulnerable will receive no compensation. While the arguments for not defending the coast are sound and sensible, it hardly seems just to say that people can lose everything as a result, and may even end up paying to demolish their own houses. If it is not financially justifiable to defend a bit of coast, then some of the money involved could be used to compensate the individual householders.

Summary of Chapter 3

The chapter has introduced aspects relating to one of the key activities at the coast, that of land claim and coastal defence. The topic is vast but represents perhaps the greatest impact that humans have had on the coast.

The tendency of policy to protect what is present, never to allow land to be given to the sea, is becoming outdated. Since we have been all too ready to take land from the sea (land claim), so we should also be as ready to consider the prospect of giving it back when natural conditions so dictate. This is not to say that wherever the sea erodes, land should be surrendered. Issues of land use need to be brought into consideration because humans have already gone so far down the line of coastal abuse that in many places to surrender land is non-viable. What is important, however, is that whenever a coastal erosion problem occurs, due consideration is given to all possible options, and that the eventual policy is considered over a coastal cell, not just on a local basis. If this means that development has to be relocated to allow sediment to be input to the system, then so be it. After all, the cost of defending may well exceed the cost of relocation.

It is for these reasons that this chapter has been presented in the way that it has. Introducing the problems and causes makes clear the role of coastal defence. Thus, the need having been established, the various approaches to defence can be presented with due acknowledgement of the fact that while there are benefits for each methodology, there are also environmental issues which need to be considered. After all, the underlying theme throughout this book is that the best way of managing the coast is to leave it alone. Hence any interference, be it development or defence, affects these natural processes.

Increasingly, the policy of leaving the coast alone, helping it to do what it wants, or providing natural habitats is becoming more common. Such processes, known as 'soft' methods (as compared to the 'hard' methods of building sea walls and groynes, etc.), are becoming more important than what have hitherto been the conventional techniques, but nevertheless the impacts of such actions still need to be considered.

Hence, the intention of Chapter 3 has been to provide the reader with details of the main aspects of defence practice and methodologies, as well as their environmental implications. The reader should leave this section with an understanding:

◆ of the reasons behind the need for defences and the role of land use and hinterland values;
◆ of the various methods available for protection, their applicability, and their impacts;
◆ of the contrasting nature between soft and hard methods – habitat creation v. solid structures;
◆ of how humans strive to minimise the impacts of defence structures through policy and the planning process;
◆ that in all cases, the one thing that needs to be considered is to leave the coast alone to 'do its own thing'.

Chapter 4

The industrial use of coasts and estuaries

Introduction

The developed world has become very much an industrially based society and there is an increasing trend for the developing world to follow the same route. Within society, large areas of land are given over to industry and its associated infrastructure, for the manufacture, processing, and storage of commodities. Perhaps before we proceed it is necessary to understand what is meant by 'industry' in this context. As far as we are concerned here, industry is a term used to refer to those aspects of society which manufacture goods or products which are beneficial to it. By using such a broad definition, we include not only manufacturing industry, but also power stations. We exclude tourism, however, as the problems for the coast of this activity are sufficiently different as to warrant separate treatment (see Chapter 5).

In order for industry to function, there are several factors related to the environment which need to be met:

1   suitable land for development. The definition of suitable land may vary and is somewhat industry specific, but often large areas of flat, cheap land are required, especially for petrochemical works, power stations, and oil refineries;
2   the supply of raw materials: basic needs to start the manufacturing process such as crude oil for refining or for petrochemicals, coal and oil for power generation, or imports of raw materials;
3   the need for water: many processes require large quantities of water, either in the actual making of products (e.g. electricity generation) or for washing and cleaning;
4   means of waste disposal: many by-products of society are discharged into the environment, although, increasingly, they are being processed and treated first. Hence, some transport pathway is needed;
5   a means of transporting the finished goods to their eventual markets.

Such requirements restrict the number of suitable environments which are available, and it is common to find certain industries concentrated in specific locations which satisfy all, or the majority of, the conditions detailed above.

The significance of coasts and estuaries for industrial location

In answering the question of why coasts and estuaries are significant in the context of industrial location, one has only to look at the five points listed above. Estuaries, in particular, provide large areas of flat land, often at cheap prices because the land is newly claimed. They are also areas for port development, again partly owing to land availability, and so the import of raw materials and the export of the finished products are relatively easy, and the potential supply of water is limitless. In addition, estuaries have also been seen, along with rivers, as natural 'waste disposal systems'; 'things' into which industry can pour its waste to get rid of it. While this has certainly been true in the historical context, more recently this situation has been reviewed and strict legislation regarding discharges has been brought into effect, so that the situation is by no means as serious as it once was. This topic will be returned to in later sections of this chapter.

It can be seen, then, that the locating of industry in an estuarine environment provides it with all the basic facilities. This fact, however, naturally puts great demands on these systems, and, as between coasts and estuaries, it is typically the estuaries which suffer the most. The main reason is that industries requiring ports or facilities for loading and unloading of imports and exports tend to locate in estuaries, because estuaries, for reasons already discussed, are where ports are best located. Table 4.1 lists industries commonly located along coasts and estuaries, and, as can be seen, the majority are estuary, rather than coast, specific.

Other industries, particularly power generation, may not need all the above-mentioned facilities. Nuclear power stations, for example, need just a good land-based infrastructure, an abundant supply of water and, commonly, a location away from large centres of population. In these respects, coasts are often as suitable as estuaries. In some cases, such as Sellafield on the Cumbrian coast of the UK, the continuous discharge of low-grade radioactive waste actually favours a coastal location for reasons of dispersal. However, from a contamination point of view, this can lead to problems of accumulation, as has occurred in the case of Sellafield, where natural longshore

TABLE 4.1   The relevance of coastal and estuarine sites for industrial developments

Industry	Estuary	Coast
Power generation	✓ (coal, gas and oil)	✓ (nuclear)
Ports	✓	×
Oil refineries	✓	×
Petrochemical works	✓	×
Oil rig servicing/storage	✓	×
Shipyards	✓	×
General manufacturing	✓	✓
Domestic waste discharge	✓	✓
Sewage disposal	✓	✓
Landfill	✓	×
Aggregate extraction	✓ (minimal)	✓
Shellfish collecting	✓	✓
Commercial fishing	✓	✓
Fish farming	✓	×

drift leads to the accumulation of nuclides in many estuaries, and especially in Morecambe Bay, where some nuclides can reach many times those of background.

As a result of these basic industrial requirements, it can be argued that it is necessary for such development to occur at the coast, and so it is necessary for such developments to be accounted for in the coastal planning process as a necessary future development. Increasingly, though, coasts and estuaries have also seen new industrial growth not due to the requirements already mentioned, but purely on the basis of cheap land. Many of these developments, such as business parks, light industry, or offices, do not need to be adjacent to the coast, and could equally be located further inland. As a result of this unnecessarily high concentration of industry around coasts and estuaries, land values have been increased, thus increasing the need for coastal defence measures where a coastline becomes threatened with flooding or erosion. Hence such large-scale development has given rise to potentially serious future coastal management problems due purely to land prices.

Another important factor in the establishment of industry is exactly where on a stretch of coastline the industry is located. Given that many industries are necessarily located at the coast, the siting of these needs to be done in sympathy with local environmental conditions. It is not, for example, a good idea to site a petrochemical works on a rapidly eroding coastline. Clearly, such a situation would never occur because of the planning process, but, considering that such an installation may have a life expectancy of fifty years, it is necessary to predict what the coast will be like in fifty years' time, a prediction which, given our limited knowledge of large parts of the coastline, is not necessarily a straightforward undertaking. A petrochemical plant, for example, built 1 km inland from the coast might, during this fifty years, find itself right on the coast if erosion starts to become a problem. This would necessitate increased coastal defence expenditure in order to protect the plant, whereas if it were sited elsewhere these problems could have been alleviated. A good example of such a problem occurs in the case of the Easington gas terminal on the Holderness coast of eastern England. This development, located at the southern end of the rapidly eroding Holderness coast, is threatened by the erosion of the soft till cliffs. When constructed, the cliff was some considerable way from the development, but, given rates of erosion measured in metres per year, it is apparent that what was considered a safe distance in the planning stage is proving not to be so. Because of the importance of the site, it will be necessary to protect this stretch of coast in years to come, unless the infrastructure can be relocated (which may be a viable option, as that located nearest to the cliff is the oldest part of the plant and nearing the end of its planned operating life). As we have seen in Chapter 3, the supply of sediment from this coast is critical to the local sediment budget, and so it becomes equally important to protect the supply of sediment from other parts of the coast, so that nearby villages will not need protection.

The problems of long-term coastal stability and predicted erosion trends are particularly critical in the siting of nuclear power stations. Such installations are invariably sited on open coasts owing to the large water volumes required. Those located in estuaries tend to require the construction of large reservoirs owing to the need to safeguard water supply at all states of the tide. In addition to its operating life, a reactor will need several centuries for radiation levels to reduce. Hence, the construction of nuclear power stations on a coastline means a long-term commitment to defence, one of the order of three hundred years. Such a long-term prediction is difficult even in situations where the behaviour of a coastline is well known. Along coasts where knowledge is sparse, predictions are almost impossible, especially given the added problem of sea level rise. As far as the UK is concerned, the siting of nuclear power stations has often occurred in the absence of an adequate appreciation of coastal processes and geology (Box 4.1). The result of this is that many nuclear installations, in view of the mobility of many of the coastlines involved, and the role of sea level rise, may require considerable protection for several centuries to come. The alternative is to remove the hazard when the power station closes. At the present time, this is not considered a practical option, as the highly radioactive core presents a major environmental hazard, and the technology to break it up and remove it is not fully established. In addition, purely on economic grounds, the cost of removal is likely to be far in excess of the cost of defence, and so, in the majority of cases, increased coastal protection is likely to be the favoured course of action. Such coastal protection, however, must happen irrespective of coastal processes, management plans, or the overall sediment budget of the coastal cell, and will therefore contravene many of the principles which we have established regarding how coastal systems should be managed. Local defence authorities must, in effect, write a blank cheque for the future protection of the land.

Another aspect of industrial development at coastal or estuarine locations is pollution. Industrial waste is discharged into the estuary, or out to sea, where dispersal will remove it from the local area. However, many of the world's coastlines have experienced major problems associated with industrial emissions. We have seen how both wave and tidal currents can move sediment onshore, offshore, and along shore. As will be seen later, this sediment may not just represent a mineral grain, but may have pollutants attached to it. Furthermore, particulate pollutants, such as mine waste or coal dust, will behave like sediment and thus will be moved around in currents. These problems will now be considered in more detail.

Historical trends in waste disposal

The exploitation of any natural resource generates waste material. The disposal problem associated with large quantities of waste generated during periods of industrial or mining activity has meant that the natural environment has been the ultimate recipient. The assumption of many industrialists that both estuaries and coasts may be regarded as natural waste disposal systems has meant that these environments in particular have suffered intense pollutant input. The original assumption, especially in the case of estuaries, is not really justified, since pollutants can remain in the system and become incorporated into the sediment record of such a system. Whenever muddy sediments are deposited, they absorb many pollutants from solution, and hence contain within them whatever levels of pollutants are present in the water body at the time of deposition. On deposition, the metals are stored within those sediments, but if

BOX 4.1 The relationship of nuclear installations and coastal stability in the UK

Because of the basic requirements of nuclear power, the coast is seen as the most suitable location for such developments. In the UK, the only inland sites tend to be those of reactors used for research; practically all the electricity generating stations are located on the coast.

The problem is that many of these installations are located in areas which have potential coastal defence problems. In September 1992, the UK government issued a guidance note on coastal planning which stated that any new developments away from urban areas should not be permitted on coasts with an erosion problem. Around the coast, there are already thousands of developments which would not have been allowed under these restrictions, including eighteen of the UK's nuclear installations which are located on land liable to flooding or erosion and are entirely dependent on coastal defences for their stability. What is perhaps worse is that some of these defences are not really adequate. We have seen already the problems with defences in the south-east of England. Constant land subsidence, sea level rise, and retreating marshes all point to a coast which is unstable and could best be managed by being allowed to retreat inland. Coincident with this area are the nuclear power stations at Bradwell in Essex, Sizewell in Suffolk, and Dungeness in Kent.

The table details the current power stations (see map for the geographical distribution), the land on which they are built, and stability of the coast. The table includes sites, but does not indicate the number of reactors on each site (indicated in parentheses on map). Sizewell, for example, has two com-missioned reactors and a further one planned – this on a site of unconsolidated shingle which is likely to be subject to erosion and increased mobility. Similarly, there are two reactors on the shingle at Dungeness, a site of histori-cally rapid accretion, but one showing increasing trends towards erosion. The site at Berkeley has been decommissioned, and will be left with the highly radioactive core encased in concrete. This site is claimed marsh, lying below mean high water, and necessitating a long-term commitment to maintaining or improving the current flood defences.

It is likely that of those sites listed in the table, numbers 5, 6, 7, 9, and 10 are likely to pose future defence threats, with 3, 4, 8, 11, and 12 also posing possible problems, depending on rates of sea level rise and increases in storminess: In reality, all these sites, and those in other countries, will need a commitment to protection for several hundred years. Nobody can predict what the coastline of a country will look like at that time – what areas will have eroded, what areas will have accreted, etc. Thus there are many unknowns. What is certain is that developing reactor sites at the coast means that whatever happens to the coast, these areas will have to remain protected, even if the land around them becomes highly erosive.

Number on map	Name	Area of construction	Coastal stability
1	Dounreay	Old red sandstone	Stable
2	Torness	Limestones	Stable
3	Hartlepool	Permo–Trias	Locally unstable
4	Billingham	Permo–Trias	Locally unstable
5	Sizewell	Shingle spit	Future erosion, shingle mobility
6	Bradwell	Claimed marsh	Inadequate defences
7	Dungeness	Shingle spit	Future erosion, shingle mobility
8	Hinkley	Jurassic bedrock	Prone to rock falls
9	Oldbury	Claimed marsh	Sea level rise, intertidal erosion
10	Berkeley	Claimed marsh	Sea level rise, intertidal erosion
11	Heysham	Limestone plus till	Stable
12	Sellafield	Till cliffs	Eroding
13	Chapelcross	Volcanic bedrock	Stable
14	Hunterston	Limestones and sandstones	Stable

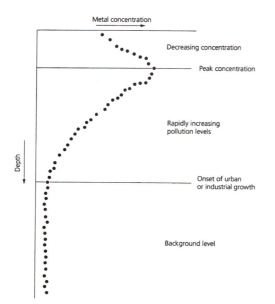

FIGURE 4.1 Diagrammatic representation of typical heavy metal concentration v. depth trends from polluted salt marsh sequences
Source: French, 1990

the sediments start to erode, then these pollutants re-enter the system. In other words, the eroding sediments will behave as a pollutant source. What this means, therefore, is that in estuaries in particular, when tidal currents move clay and fine silt (mud), they also move pollutants with it.

The large-scale production of pollution became of significance only around the early to mid-nineteenth century during the period of sustained industrial and economic growth commonly referred to as the Industrial Revolution. At a point in a sediment sequence coincident with the onset of this historical period, sediment-borne pollution levels show a rapid concentration increase (Figure 4.1). However, this is not to say that this was the start of industrial activity in a particular area, as localised industry and mining adjacent to some estuaries can be documented right back to the Romano-British period, normally associated with natural exposures, where collection of coal or metal ore could proceed without technical difficulty. Swansea, in the Bristol Channel in the UK, for example, had been an important centre for copper smelting in the seventeenth and eighteenth centuries (Flinn, 1984).

To a large extent, the amount of pollution entering the environment is positively correlated with the growth and decline of towns and industry. Areas around estuaries are, as we have seen, especially favoured for the development of industry. The Severn estuary/Bristol Channel system, along with its extensive catchment, is one of the UK's most heavily industrialised estuary systems (Price, 1986). On its northern margin, there has been industrial and mining activity in south Wales and the Forest of Dean dating from the Romano-British period (Hamilton *et al.*, 1979; British Coal, 1986). Until the mid-nineteenth century, however, activity was only localised. Following the Industrial Revolution, there was a rapid increase in industrial and mining activity as the railways opened up the large south Wales coalfield for development. In addition, the railways provided a link to the coast, allowing the development of a coal export trade through the rapidly expanding ports. On its southern margin, there is the large chemical and processing complex of Avonmouth and Severnside, in operation since the early 1900s, and also, prior to 1973 (British Coal, 1986), the coal-producing area of the Bristol and Somerset coalfield. The Severn catchment includes areas of the south-west Midlands, where industrialisation and urbanisation have followed the same patterns of development as areas adjacent to the estuary. From these industrial centres, pollutants can be transported to the Severn estuary and Bristol Channel along a variety of pathways (Figure 4.2), the relative importance of which, with specific reference to the Severn, can be seen in

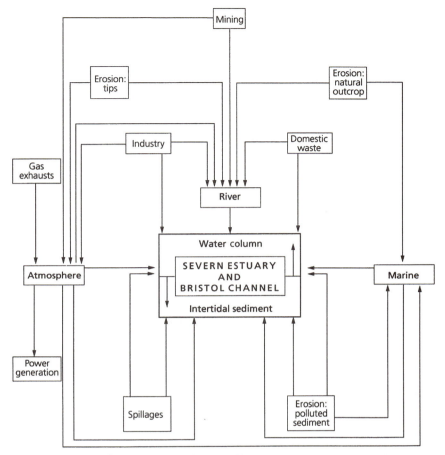

FIGURE 4.2   The Severn estuary and Bristol Channel system as an example of the causes of estuarine and coastal pollution, and the various transport pathways
Source: French, 1990

Table 4.2 (Owens, 1984). Although this example relates specifically to the Severn estuary, the principal pathways shown in Figure 4.2 and outcrops are exactly the same for any industrial catchment, and so it is possible to substitute the name of any such estuary into the central box to achieve a flow diagram relevant to any system.

The most direct way for pollutants to enter an estuary or coastal system is via direct discharge; that is, through a pipeline from the factory into the channel. This is not, however, the only means. Figure 4.2 details all the various pathways, and demonstrates routes for land, river, and the atmosphere. It can also be the case that different pollutants predominate in the different transport media such as air or water, and also, the importance of these media with respect to different pollutants may vary within single systems. Using the Severn estuary as an example, it can be seen from Table 4.2 that lead and zinc are chiefly derived from atmospheric input, with industry

TABLE 4.2   The relative importance of different transport media in the supply of metal pollution in the Severn estuary

Metal	Rivers and streams[a]	Atmosphere	Other[b]	Industry	Sewage	Sludge	Total input (kg day^{-1})
Mercury	42.2	0.0	57.8	54.4	2.2	1.2	13.3
Cadmium	32.4	21.1	46.5	32.6	12.5	1.4	61.6
Lead	16.9	57.8	25.3	18.5	4.5	1.9	1,150.0
Zinc	30.3	42.2	27.5	17.8	8.8	0.9	5,160.0
Copper	54.2	19.0	26.8	15.8	8.6	2.4	621.0
Iron	68.9	5.8	25.3	17.2	8.1	0.0	62,700.0
Manganese	72.7	5.9	21.4	16.2	6.0	0.0	2,780.0
Nickel	75.9	11.9	12.2	3.3	7.3	1.6	464.0

Source: Modified from Owens, 1984

Notes: [a] Includes industrial, storage/tip/natural erosion.

[b] Includes industry, sewage, and sludge components.

Figures are quoted as percentage of total input, unless otherwise indicated.

and rivers being of secondary importance. Atmospheric lead mainly results from the large quantities found in vehicle exhaust fumes, while atmospheric zinc is chiefly derived from non-ferrous metal production (Pacyna and Münch, 1987). Significant quantities of both lead and zinc are also present in the gaseous emissions from coal- and oil-fired power stations (Pacyna and Münch, 1987).

In this example, the derivation of atmospheric-borne pollutants (Table 4.2) is likely to be of relatively local origin. With the overall prevailing wind direction being from the west or south-west, the majority of wind patterns would have developed out in the Atlantic Ocean, where there are no pollutant sources. The first real sources of airborne pollutants are the industrialised areas of the Welsh and Avon coastlines. The second most frequent wind direction, from the north-east, may derive airborne pollutants from the area around the south-western Midlands. With attention turning towards a greater concern for the environment, atmospheric inputs may well be expected to decline as lead-free petrol increases in popularity, and as tighter controls on air pollution come into force. Thus it can be seen that the potential to supply pollutants is great in many areas, but that the method of transport and sources will vary. In our case, however, these factors are perhaps not of prime importance, although the study of the role of pollution in estuaries and on coasts does need to take source and transport into account. What is of great significance, however, is the fact that the pollution ends up in estuarine and coastal environments, and that once there it can become incorporated into sediments and so play an important future role in pollutant sources should these sediments begin to erode.

The historical development of many areas has meant that the supply of pollutants to coastal and estuarine sediments has a simple temporal variation (Figure 4.1). The level of most pollutants is dependant on two factors. First, there is a background component due to natural processes of erosion and weathering of outcrop

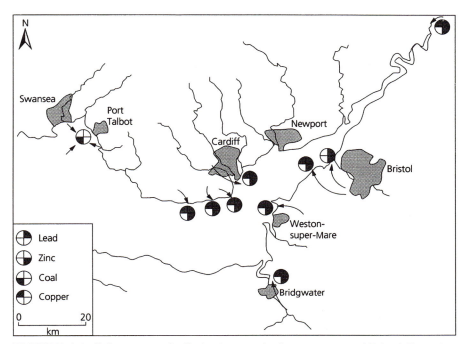

FIGURE 4.3   Point sources of pollution input to the Severn estuary and Bristol Channel
Source: French, 1990

within the catchment. Second, and superimposed on this, are the levels of pollutants, variable in time and space, associated with anthropogenic activity. Some pollutants do not occur naturally and so cannot possess a component due to background levels, in which case their presence is purely a result of anthropogenic inputs. It is these anthropogenic inputs which have the greatest temporal variability, and which have been used to form the basis of a pollutant-based chronology (Allen, 1987, 1988; French, 1996a, b). In addition, some spatial variation also arises from the non-uniform way that pollutants are introduced into the system (French, 1993a). Industrial discharges produce a series of point sources from which pollutants are disseminated by the system to become diluted and mixed. This mixing is variable in nature but is primarily a function of the mixing capabilities of the system. Such a model of point sources can be seen in Figure 4.3, which represents four of the principal pollutants in the Severn estuary and Bristol Channel, all of which are linked to industrial or urban areas.

The increase in the amount of pollution due to anthropogenic influence can be better understood when compared to the levels associated with natural erosion. Jones (1978) exemplifies this point in tabular form (Table 4.3). Whereas some metals (e.g. silver and nickel) show little increase due to anthropogenic influence, others are dramatically affected, with up to 110 times natural levels in the case of tin.

It is clear that many estuarine sediments, therefore, contain within them stores of potential future pollutants. This raises a serious issue as to how much pollution is

TABLE 4.3   A comparison between natural and anthropogenic rates of heavy metal input

Metal	Rate of natural erosion (N) (10^3 tonnes yr^{-1})	Rate of anthropogenic supply (A) (10^3 tonnes yr^{-1})	A/N
Tin	1.5	166.0	110.0
Antimony	1.3	40.0	31.0
Lead	180.0	2,330.0	13.0
Iron	25,000.0	319,000.0	13.0
Copper	375.0	4,460.0	12.0
Zinc	370.0	3,930.0	11.0
Molybdenum	13.0	57.0	4.4
Manganese	440.0	1,600.0	3.6
Mercury	3.0	7.0	2.3
Silver	5.0	7.0	1.4
Nickel	300.0	358.0	1.2

Source: Modified from Jones, 1978

present in these stores, and what the future threat is likely to be. To date, much work has looked at levels present in sediments, but hardly any has examined the linking of these to rates of erosion, and hence the future role of supply. Many reports on pollutant variation with depth have been utilised for the benefit of increased understanding of sediment accretion rates, resulting in the development of pollution-based stratigraphies. Many of these are centred around sites from the USA. Finney and Huh (1989) analysed a series of box cores from Santa Monica and the San Pedro Basin, California, for a series of metals and for lead-210 dates. From this they showed that variation in copper, lead, and zinc levels arises in response to anthropogenic influences, and that it reached a peak level around 1970. Since this peak, levels have been slowly declining to the present day, where they are now around a third of this peak level.

Other reported findings (Hamilton-Taylor, 1979; Crecelius, 1975) all tend to show the same pollutant/depth trend as the general pattern shown in Figure 4.1. Förstner and Müller's (1973) review of metal pollution in rivers in response to anthropogenic influences also supports these ideas. The recent study of Middleton and Grant (1990) looks at a range of heavy metals in the Humber estuary and relates these to a pre-industrial datum horizon (Scrobicularia Clay). Their data, as a result, use this comparison to enable it to be expressed in the form of a series of enrichment factors which show the impact of discharge on coastal sediments. A similar demonstration was also produced for the Severn estuary (French, 1993b, 1996b), and clearly demonstrates the impacts of industrial development in this system. Other examples are detailed by Allen (1987, 1988), Allen and Rae (1986), de Souza *et al.* (1980), French (1990, 1993a, b, 1996b), Hamilton and Clarke (1985), Itoh *et al.* (1987), McCaffrey and Thompson (1980) and Müller (1981). One of the most striking features to arise from all these sets of data is the remarkable similarity in the form of the pollutant–depth trend (see Figure 4.1). In most cases, the results show a steady, low background level which, at a point governed by the history of the individual catchments, begins to increase rapidly to a peak level of pollution, before slowly declining to the present day.

Despite the predominance of studies on metals, these are not the only form of pollutant found in estuarine and coastal systems. A second major group concerns the particulate pollutants. However, studies on particulate pollutants are rather more scarce than those relating to heavy metals. The majority of the work is concerned with areal surveys of present-day environments. Hunt *et al.* (1984) looked at heavy metals associated with particulate atmospheric pollutants, and developed a link between metals and magnetic mineral concentrations in a variety of environmental contexts. By characterising the magnetic components of a variety of emissions, they were able to suggest an identity for emissions of unknown origin. Magnetic susceptibility is also the method used by Oldfield and Scoullos (1984) in particulate monitoring in the Elefis Gulf, Greece. They showed that magnetic susceptibility could be used to characterize sediment samples and filtered particulates.

Studies by Campbell *et al.* (1988) link the two fields of adsorbed and particulate metals. Using the Mersey estuary in north-west England as an example, they studied the contrasting behaviour of dissolved and particulate nickel and zinc, using this work to suggest origins for the two metals. A combined study of dissolved and particulate trace elements comprised the study of Abdullah and Royle (1974) in the Bristol Channel. Two surveys carried out in 1971 show that the major source of these metals is run-off from areas of mineralisation and waste disposal.

It is clear that there has been much work done on the levels of pollution present in coastal and estuarine systems. Whereas the absolute concentrations may vary, the temporal trends are consistent with the variation in industrial and urban activity. It is true to say that many of these reports have been determined from a purely academic standpoint. There are, however, many implications for the estuary and coastal environments which arise out of this. What all these have in common is that they all demonstrate how human activities, via industrial development and urbanisation, have produced large-scale impacts in estuaries and along coasts. While visually these may not be readily noticeable, this potential with respect to future pollution problems is certainly real and may have serious consequences.

All the studies mentioned above are based on the fact that when sediment is deposited, it contains within it the levels of pollution present in the water environment at that time. An important question can now be raised. What happens to this pollution after it has been deposited? Clearly, once a marsh, intertidal mudflat, or other contaminated sediment body starts to erode, these pollutants will be reintroduced into the water body. Hence the sediments themselves become a source of pollutants in exactly the same way as a factory outfall would, with the main exception being that salt marshes do not come under the strict legislation concerning levels of emissions.

A second possibility is that even before erosion, material can be leached out of a sediment sequence by some chemical process. Under the right environmental conditions, decay of organic matter, oxidation/reduction conditions, as well as groundwater chemistry can cause the removal of metals. Such processes are complicated, and the exact nature of the process does not concern us greatly. Moreover, they are of considerably less importance than erosion.

We have a situation, therefore, where many of the muddy deposits of the industrialised world contain pollution. To put this another way, as sand dunes and salt marshes can be seen as sediment stores, all the world's industrialised estuaries can also be seen as pollutant stores. These pollutants can be adsorbed onto sediment grains or organic material, or occur in particulate form. It is important that the role of these pollutants be assessed for the impact which they can potentially produce.

Effects of pollution on estuaries and coasts

The incorporation of pollution into sediments can occur in many ways. For our purposes it is not necessary to go into great detail on all of these. Particulates will behave like sediment grains, with deposition depending on size and specific gravity of the particle. Metals, on the other hand, are adsorbed directly onto particles. This adsorption is very much grain-size dependent, as a given volume of clay-sized grains will have a greater surface area than the same volume of sand grains. As a result, for two estuaries with exactly the same environmental variables and metal pollution levels, but one sand-dominated and the other mud-dominated, the levels of metals in sediments will be higher in the mud-dominated estuary than in the sand-dominated one, purely as a result of the effect of grain size.

The adsorption of heavy metals onto clay minerals, organic material, or iron/manganese oxide/hydroxide grain coatings has been recognised as an important controlling process with respect to the deposition of heavy metals within the estuarine environment (Jones, 1978). Clay minerals, owing to their relatively large adsorption capacities and their tendency to accumulate in many estuaries, are important in this regard, and it is for these reasons that the greater part of the heavy metal pollution occurs in association with the finer sediment fraction and hence is a particular problem in muddy environments, such as estuaries. The literature concerning this topic is both vast and diverse, a large proportion being cited in papers by Allen (1987, 1988), Allen and Rae (1986), Förstner (1989), Krumgalz *et al.* (1990), and Ridgeway and Price (1987). Following adsorption onto sediment grains, in order that the pollutants can be incorporated into the sedimentary record they have to be deposited along with the sediments of the system. Sedimentation processes in estuaries and coastal environments, where most of these pollutants end up, are complex interactions between many environmental variables, notably current velocities, water chemistry, water circulation and sediment grain size (see Chapter 2).

Types of pollutants

Many substances which occur as pollutants, or are indicative of pollution, build up in the natural environment, a process referred to as 'geo-accumulation' (Müller, 1981).

Pollutants can generally be regarded as occurring in two principal forms. First, there are chemical solutions or colloidal suspensions which contain heavy metals and other chemical compounds. Such material can occur as a result of natural processes – that is, chemical weathering, etc. – or by the discharge of waste during anthropogenic activity. Second, there are particulate pollutants: solid material such as wastes from mining activity or from power generation (fly ash, spherules, etc.). All these materials, both heavy metals and particulates, occur in discharges and all, therefore, have the potential for being incorporated into the sedimentary record of a system. By using dated sediment cores, it is possible to study the temporal importance of many of these substances over the past century or so. Müller (1981) has produced a generalised chronology of various groups of pollutants with respect to temporal variation in environmental concentration (Figure 4.4). The following classification is based on Müller's work.

I have said that pollutants can be divided into two main groups, but this is perhaps a little simplistic, because not all chemical solutions behave in the same way.

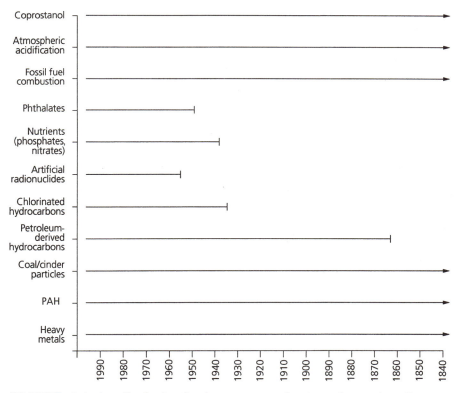

FIGURE 4.4 Age distribution for the occurrence of major pollutants in sedimentary deposits, based on the work of Müller (1981)
PAH = polycyclic aromatic hydrocarbons

framework is that adopted by Müller (1981), which divides pollutants into eleven groups (Figure 4.4), each of which has its own set of time-related properties. The evolution of *heavy metal* concentration runs parallel to world-wide industrialisation, which began around the mid-nineteenth century. A significant proportion of the increase has been attributed by Erlenkeuser *et al.* (1974) to the combustion of coal and lignite. *Polycyclic aromatic hydrocarbons* (PAH) show a strong positive correlation with heavy metals. PAH also started to increase around the latter half of the nineteenth century, reaching a maximum in the mid-1960s. This close correlation has led to the conclusion that coal combustion is an important source not only of metals, but also of PAH, because, as Müller states, 'it is a known fact that the incomplete combustion of coal produces PAH'. *Coal particles* have an overall similar evolutionary pattern to that of metals and PAH, but are more strongly influenced by local variations. Coal-fired boilers and coal production are important sources. *Petroleum-derived hydrocarbons* from crude oil or refined petroleum first appear around 1870, although the largest growth is in the late 1930s in response to a rapid increase in vehicular traffic. *Chlorinated hydrocarbons* have become significant pollutants since the 1940s. Since 1970, owing to their high toxicity and carcinogenic properties, their application to a closed system, such as a lake, has been legally restricted, although a decrease in levels is yet to be revealed by the sediment record. *Nutrients*, particularly phosphorus and nitrates, reflect the trophic (production) status of an aquatic system. *Artificial radionuclides* appear in the mid-1950s in response to the atmospheric testing of nuclear weapons. Since 1963, levels have been decreasing as strict legal constraints have been implemented. In addition, nuclear reactors and reprocessing contribute to the sediment-borne radionuclides. The accident at Chernobyl, for example, produced enough caesium-137, and the characteristic caesium-134, to deposit a unique layer in many sediment environments (Rowan *et al.*, 1993). This layer can be detected and forms a unique reference horizon from which to date post-Chernobyl sediment accumulation. *Phthalate esters* are widely used as plasticisers in the manufacture of PVC and other plastics. Concentration in sediments has increased steadily since the 1950s in response to an increased demand for plastic products. *Coprostanol* (faecal pollution) is recorded in sediments back to pre-industrial times and correlates with population growth.

These principal groups represent a convenient system in which to categorise pollutants and to study their temporal importance. Perhaps, in comparison to some work, the classification may still be a little crude; for instance, the inclusion of coal and cinder into one group brings together a diverse range of material which might be better subdivided. However, as far as we are concerned, the impact of different pollutants on the environment varies with time. Eroding marshes, for example, will not reintroduce pollutants if they were deposited before the input of that pollutant started. Hence, we may expect older marshes to contain a pollutant such as coal or heavy metal, while newer ones may, in addition, contain pollutants such as chlorinated hydrocarbons or radionuclides.

Effects on the system

We have seen how release of pollutants into the environment can be reflected in the uptake of these substances by sediments. These inputs can impact on the environment in two ways. First, they can cause pollution when they are initially pumped into the water. Second, they can also produce increased pollution inputs when the sediments in which they are contained start to erode. There is, however, another factor, which may be seen by some environmentalists as a positive issue. When pollutants are absorbed onto sediments, they are being removed from the water body, so that the water quality becomes better. Hence the removal of pollution by sediment adsorption can be seen as a water-cleaning process and, as such, of clear benefit to a heavily polluted environment. It is important to remember, however, that these processes are not permanent, and that subsequent re-release is likely in the future since pollutants are not being removed from the system, merely stored away from the water body temporarily. In systems where pollution has been high, which has resulted in high levels in sediments, such as occurs in some shipbuilding areas for example, it may be environmentally preferable to physically remove the sediments from the environment to be treated elsewhere. Fraser (1993) reports on the case of New Bedford harbour, Massachusetts, where heavily contaminated sediments containing polychlorinated biphenyls (PCBs), heavy metals and other contaminants caused the harbour to be placed on the National Waste List as one of the USA's most hazardous waste sites. Various forms of dredging were employed to remove the sediments from the harbour, and these were then disposed of in one of two ways. First, they were placed in confined disposal facilities adjacent to the estuary, where sediment was allowed to settle in sediment ponds before burial. Second, contained aquatic disposal was used, where contaminated material was sealed in holes dug in the floor of the upper estuary. While from a pollution point of view this form of treatment would be sensible, it should be remembered that the process involves taking much sediment from the system, and hence has a negative effect on the sediment budget.

The impact which these pollutants have on the environment can increase or decrease according to several environmental parameters. The primary impact will be determined by the concentration of the initial input. On emission, the concentration will be much greater around the source, which will gradually decrease as the material disperses from this source. A coastline with strong onshore or offshore currents will spread the pollutants over great distances. In contrast, such currents can also be responsible for washing material onto beaches. In estuaries, the flood and ebb tides are largely responsible for the dispersion of material, with stronger currents producing a greater degree of mixing. The estuaries with the stronger currents can achieve an almost uniform mixing, with pollutants dispersed to produce uniform levels throughout the system. Because ebb and flood tides introduce material into, and take it out of, an estuary, over time pollutants will eventually leave the system and be taken out to sea. The time taken from introduction to removal is known as the flushing time. Flushing times vary considerably and can range from a few days to many months. The significant fact, however, is that estuaries with long flushing times will retain pollution in the system for longer, and thus sediments will have a longer period of time in which to adsorb material. It can be argued, therefore, that there will be greater retention of material in estuaries with long flushing times.

It is still possible for metals, once deposited, to re-enter the system, as mentioned earlier. This potential is related to metal stability (Förstner, 1989). On deposition, a layer of sediment represents a single time plane which will contain within it levels of pollution governed by the pollution state of the estuary at that time. One important assumption is that when deposited, the pollution record is permanent and stable, and that no significant post-depositional migration of the heavy metals occurs. Certainly, in the case of anoxic, subtidal muds, problems of metal mobility do occur, with a tendency for metals to combine to form sulphides. In some cases, these metals then become mobilised, with the result that they can migrate into interstitial pore waters and up to the sediment surface on a concentration gradient set up by the overlying water body (Elderfield and Hepworth, 1975). In intertidal areas, such as high mudflats and salt marshes, prolonged anoxia is uncommon and water cover is not of a long enough duration to allow concentration gradients to be established. As a result, it is thought (Rae, pers. comm.) that no large-scale mobilisation occurs. Work by Allen *et al.* (1991), based on a Severn estuary salt marsh sequence, suggests that in this environment at least, post-depositional mobilisation occurs only from the organic phase – that is, metals adsorbed onto organic material – with other phases being stable. As organic matter decays, metals are released back into the water body. There is no evidence to suggest that once released, these metals are readsorbed by another phase. The rate of release (Allen, pers. comm.) is estimated as being in the order of a 50 per cent decrease every few decades. If metals did migrate up a salt marsh sequence on a large scale, then one would expect to find a gradual up-section increase in metal levels, with the greatest concentrations at, or just below, the surface. In other environments, the mobilisation of particle-bound material (i.e. heavy metals adsorbed onto clay grains or oxide/hydroxide coatings) can be enhanced by five major factors (Förstner, 1989):

1   Acidification can be a prime factor in mobilising metals in water bodies and groundwater, and in the production of acidic drainage from coal and ore mines.

2   Salinity increases can be critical in the remobilisation of metals from cadmium-rich sediments in estuaries (Salomons and Förstner, 1984), which, in conjunction with sediment-borne biota, can cause sediments to become major sources of dissolved metals.

3   Complexing can significantly affect metal mobility, but is dependent on such factors as acidity, mode of metal occurrence, and cation presence.

4   Organic activity may also affect acidity, and cause sulphide reduction and bioturbate sediments, which can cause other processes to become more prolific.

5   Oxidation and reduction processes can cause sulphur complexing, converting the metal to its metal sulphide, thus changing its stability. In oxygenating conditions, metals may revert from their sulphides to become carbonates, oxides, etc., again changing their stability.

Thus, irrespective of the erosional state of the sediments, it is still possible, albeit under specific environmental conditions, for metals to be input to the environment from sediments without any erosion.

Other implications of sediment-borne pollutants: sediment dating

The interrelationship of sediments in any estuary or coastal environment can reveal a complex series of events which have helped shape the morphology of that particular area. Stepped salt marshes or cross-cutting dune systems suggest distinct erosion and accretion periods, events which need to be put into a temporal framework in order to understand how natural processes have shaped the system. The ideas of lithostratigraphy, biostratigraphy, etc. are well founded in the geological record and, to a certain extent, in the modern sediment record, and have frequently been used to determine sedimentary events. In the study of contemporary sediments, however, the development of a chronostratigraphy (a scheme for dating and correlation) can be of more use than biologically or lithologically based counterparts. Because these sediments are deposited in environments in which humans have a direct influence, pollutants are commonplace. In effect, this means that owing to anthropogenic factors, temporally variable quantities of pollutants are often incorporated into the sediments of a system, and it is possible to use these temporally variable discharges to discover more about how a sediment sequence operates, and how its morphology is determined.

Although much work has been done regarding the pollutant status of many estuaries, the main thrust of much of this investigation has been towards looking at modern pollution levels and the possible effects that these may have on life and environmental issues in general. This means that most of the literature concerns the present-day areal distribution of pollutants and not their use in a stratigraphical context. To date, few authors have taken a stratigraphic approach, although among those who have, concentration has been focused on heavy metals.

Because of the deposition of pollutants along with sediments, a sediment core or salt marsh section will contain within it a historical record of the sediment activity and pollution history of the system which deposited it. The individual sediment layers can be dated using appropriate radioisotopes (typically ^{210}Pb or ^{137}Cs) in order to document the history of contamination (Warren, 1981). On this basis, Bruland *et al.* (1974) found that lead and zinc in coastal sediments near Los Angeles dramatically increased around 1940 along with rapid industrial development; Crecelius (1975) documented an increase in arsenic levels in Lake Washington around 1890 coinciding with the opening of a new smelter in the area; Hamilton-Taylor (1979) showed increases in lead and zinc levels in Lake Windermere sediments after 1850 following a rapid population increase; and Allen (1987, 1988) and French (1990, 1993b, 1996a, b) have shown how increased metal concentrations followed the industrialisation of the Severn catchment around the mid-nineteenth century. Although not all the cited work refers to coasts, the techniques and findings are all relevant to the situation under discussion. The analysis of metal trends can demonstrate variation in sedimentation rates, pollution input, amount of pollution stored, etc., which are all factors which need to be determined if pollution inventories are ever to be determined for estuaries as part of management plans.

Bio-accumulation

We have seen how pollution can be incorporated into sedimentary deposits and stored. In contrast to this, accumulation of metals and other aqueous pollutants can also occur in the organisms which live in the system, and this accumulation can become accentuated up the food chain as predator eats prey. Such bio-accumulation may have considerable impact on the human population because we are often part of the same food chain, generally occupying the role of chief predator.

While metals such as copper, iron, nickel and zinc are essential in minor amounts to many forms of life, in any great concentration they rapidly become toxic. One particular feature of aquatic life is that many organisms can absorb metals into their body tissues, increasing this abundance up the food chain when eaten. There have been many instances of mortality due to toxic poisoning, including transfer to humans.

In Japan, the release of cadmium from lead and zinc smelters over a prolonged period of time led to an accumulation in local fish. The eating of this fish by the human population produced a painful disease of the joints (the name of which, when translated, produces the rather appropriate name 'ouch-ouch'), which resulted in the death of 100 people and the paralysis of many others. A further example, also from Japan, occurred between 1953 and 1956. Waste containing mercury(II) sulphate was discharged into Minimata Bay and was subsequently absorbed by fish, which converted it into organic methyl mercury. These same fish form the basic diet of the local population, and as a result of the accumulation of methyl mercury, Minimata disease (mercury poisoning) killed forty-three people and caused severe disablement in sixty-eight others (Clark, 1992).

As the above example illustrates, mercury is particularly harmful to humans. Evidence indicates that around many of the world's coastlines, levels of mercury are increasing, within both sediments and fish. Many other metals have similar effects and impacts on human populations, and all of them continue to enter the environment by many routes, including domestic sewage, industrial wastes, car exhausts, and storm drains.

It is not just human populations which suffer from pollutant accumulation. An example of the direct effect that the dumping of pollutants can have is the case of the common seal and the effect of PCBs in both the North Sea and the Baltic (Clark, 1992). PCBs are organic molecules which are fat soluble, and therefore readily absorbed into body tissues. Around the Dutch Waddensee, between 1950 and 1975, numbers of the common seal declined from 3,000 to 500 although there were no telltale signs of harm; that is, mortality levels were not abnormally high, and so there was no increase in the death rate of seals. On examination of the body tissues, the seals were found to have typical levels of heavy metals but elevated levels of PCBs. Experiments showed that seals which fed on fish heavily contaminated with PCBs experienced a dramatic reduction in birth rate, which meant that the population was not replenishing itself and that this was the cause of the seal population decline. It should also be remembered that during this same period, these same fish which were contaminated were being caught and sold in shops in countries which fish these areas. A second example, tributyl tin (TBT), used as an anti-fouling paint on ship hulls, has been shown to influence the reproductive capacity of whelks, producing imposex, a condition in which females experience the development of male characteristics (Clark, 1992).

Both these examples come from the North Sea, which is an area of intensively fished and polluted water. This largely stems from the fact that the North Sea is bounded by many highly industrialised countries, and receives much of their river discharge. The exact severity of the problem is difficult to determine, but evidence has shown that the environment of the North Sea has been significantly changed over the present century as a direct result of the interaction of humans with it. Peter Wilkinson of Greenpeace argued that by 1996 the North Sea would become biologically dead because of the continued use of it as a rubbish tip for our toxic waste. This has not actually occurred, but really only as the result of intervention to clean up the area and discharges going into it. International panels have been set up to investigate this problem, and while these panels have highlighted official concern, the problem is not seen as being as severe as is made out by some conservation groups. Nevertheless, the North Sea remains one of the dirtier seas of the world.

◆　　◆　　◆

Another type of pollutant which can readily accumulate in the food chain is pesticides. Pesticides come in many forms, both organic and inorganic. Each can have a particular effect on the natural environment beyond its designated role. The most hazardous are the organochlorides, because they are both stable and environmentally persistent. One such example is DDT, which can remain effective for up to twenty-five years. Work by Woodwell *et al.* (1967), based on a salt marsh in the eastern USA, demonstrated how the bio-accumulation of DDT can operate at different levels of the food chain (Figure 4.5). This is one example of how pesticides increase in concentration up through the estuarine food chain, and have then been transferred into humans, with dangerous levels having been reported in various human tissues and fluids, notably mother's milk.

Effects of power generation

Conventional power

I have already mentioned reasons for the siting of power stations at coastal and estuarine locations, such as the need for port facilities and deep-water berths for the import of raw materials, the requirement of large quantities of cooling water, and the need for large quantities of cheap land. These are the basic requirements which mean that the coast is well suited for this type of development. There are, however, a series of factors that cause other forms of environmental problem which need to be considered in relation to detrimental impacts.

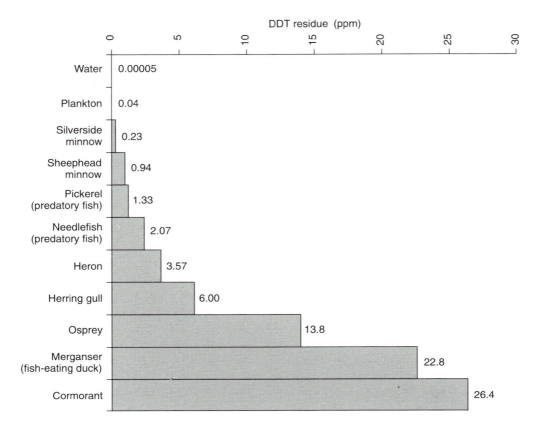

FIGURE 4.5   Accumulation of DDT up an estuarine food chain
Source: Based on data from Woodwell *et al.*, 1967

The use of water in cooling and in the generation of steam to drive turbines is common in all forms of conventional power generation. Although this water is replaced to the environment, it is at a higher temperature than when it was removed, thus producing thermal pollution. The effects of such forms of pollution are localised. Fish stocks around these warm-water plumes tend to grow more quickly than do others in colder water. While this may benefit the fishermen, it does mean that with mesh sizes effectively controlling the catch, younger and younger fish are being caught owing to the larger size, and thus fewer fish may be reaching the age of breeding maturity. As a result, the population could go into decline. It is also possible for there to be abnormal development of species, such as the development of a coral reef near Shoreham power station in East Sussex in southern England, cited earlier.

Discharges of warm water will cause localised problems. Much greater problems come from discharges of other forms of contaminant. In particular, nuclear discharges can be carried long distances via longshore drift, and also become incorporated into sediments in the same way that metals do, forming a temporal record. The majority of nuclear installations emit little, if any, radioactivity. Indeed, figures reveal that you are likely to receive a greater radiation dose from standing near a coal-fired power station than from standing near a nuclear one of similar size. However, in 1955, the scientific community was informed about a 'deliberate experiment' in Cumbria, north-west England, to study the effects of discharging radioactive effluent into the Irish Sea. Elsewhere in the world, such actions were frowned upon, and more cautious approaches were taken to such experiments.

From the onset of reprocessing in the early 1950s until the building of a new plant in 1964, discharges increased steadily. The 1970s saw the greatest increase, but it was not until 1975 that serious concern about the environment began. The Windscale inquiry (Windscale being the name of the installation at the time) of 1977 increased this worry, when the question of the effects on the local population was raised.

Much of the radioactivity emitted becomes bound to sediment and tends to be deposited offshore, where it can be ingested by benthos and, subsequently, higher levels of the food chain. Around 5 per cent of the plutonium, along with caesium-137, tends to remain mobile and becomes dispersed within marine currents, much of it ending up being concentrated in the restricted embayment of Morecambe Bay to the south. When the levels found in any environment are reported, they are quoted as amounts compared to fallout from weapons testing. These levels from fallout themselves caused great concern and criticism, but in contrast, because of concentration and sediment-bound deposition, Morecambe Bay receives around 3,000 times such levels.

Sellafield, as it is currently known, continues to discharge low-level liquid waste. Similar plants around the world have achieved almost zero discharge, using technology which was available when much of the current Sellafield plant was built. In addition, the price of such technology would not have added much to construction costs. In reality, the reason that such anti-emission equipment was not used is a management decision to discharge such substances, with a full knowledge of their toxicity and environmental persistence. As a result, many sediments contain safe, but unnecessary, levels of radionuclides.

Alternative power sources

Increasingly, scientists are turning to alternative ways to generate power, owing to the increasing threat of global warming, and the emissions of detrimental greenhouse gases which typify the burning of fossil fuels. As far as the coasts and oceans are concerned, there are two limitless energy sources: the tides and the waves. The harnessing of this potential resides with technology, much of which has not been fully proven in the production situation. Tidal power relies on the ponding up of water until there is a suitable head. This is generally done by the construction of barrages or barriers. Wave power necessitates the transfer of the energy in the waves into some mechanism for converting this to electricity. Hence, as the expression of the wave form is largest in the surface waters, the machinery may often be floated on the surface (although other techniques do exist), so that large-scale construction is not always required.

Barriers and barrages can serve one of two functions (or in some cases, both). First, some are designed to protect low-lying coasts from flooding, often following the extensive reclamation of an area, which has subsequently undergone subsidence. In this case, barrages can be raised or lowered according to conditions, as in the case of the Thames Barrier, and so are in place only when flooding is predicted. Second, barrages can be used to harness the energy of the tides to produce power. In this case, they are permanent structures which block the channel.

Flood defence (storm surge) barriers are used solely for protection against coastal flooding, and block off flow only when flooding threatens. Their advantage is that they stop the water by blocking flow, and so they can replace the need for many kilometres of flood embankments, which can have significant ecological benefits when considering natural habitats and their freedom to migrate inland. In areas of high marsh sensitivity, this can be an important consideration because the construction of flood embankments causes damage to the marsh surface. The very nature of the structure means that it can be expensive. However, when the cost is matched against the building, upkeep, and maintenance of flood embankments around the whole of an estuary, this can mean that the cost becomes more realistic.

Permanent barriers for combined flood defence and power generation cover the same uses as those built purely for flood defence, but in addition they are used as a means of power generation. On the flood tide, the barrier is open to allow water to flood into the estuary. As the tide turns, the barrier is closed to pond the water up. As the water level approaches low water, the water is allowed to ebb through the turbines, thus generating power.

Irrespective of type, both cause similar problems in an estuary. We have seen enough evidence to indicate that whenever a structure is built along the coast or in an estuary, there is potential for environmental modification and the generation of a new set of management problems. Tidal power is cited as one of the best ways of generating environmentally friendly electricity, mainly because it does not contribute to the levels of greenhouse gases (except in the construction processes). In addition, tidal power is inexhaustible but intermittent, though totally predictable and reliable. The greater the tidal range, the better the conditions for power generation. Hence, when operational, a tidal barrage can produce a continuous supply of electricity. The European Union estimates that the UK has the potential to generate 53 million kW

TABLE 4.4   The costs of developing and utilising tidal power in some UK estuaries

Estuary	Construction cost (£)	Power output (million kW)	Running cost (£ million p.a.)
Severn barrage	10.2 billion	17.0	86.0
Mersey barrage	966 million	1.4	18.0
Conwy barrage	72.5 million		0.6

Source: Based on data from Pickering and Owen, 1994

(20 per cent) of its energy demands, 90 per cent being from only eight estuaries, including the Humber, Severn, Mersey, Wye, and Conwy (Middleton, 1995). Despite this, there are no barrages in the UK, and only a few world-wide. The reasons for this are varied, but include such factors as cost and environmental impact. Table 4.4 details the costs involved in three potential barrage sites. Apart from the problems of cost, the environmental impacts are also great and include such factors as decreased intertidal areas, increased erosion, increased deposition, pollution, modification of river flow, and loss of habitat. These impacts will now be discussed in more detail.

Environmental impacts of barrages

Barrages are perhaps the most visually intrusive means of flood protection. Leaving this aside, they also have a tremendous impact on estuarine processes. These problems become increasingly significant when you consider that in the UK alone, there are practically no major estuaries without proposals for a barrage of some sort. Many minor systems have proposals, or structures already in place, which serve to pond up water for aesthetic reasons and for marina developments. We know that any structure built in an estuary is going to have an effect on tidal flow. Barriers need piers and jetties, and while these are built in such a way as to minimise intrusion, they will still have an effect. The effect of pier construction can alter the direction of tidal currents, as well as providing lee areas where sediment can accumulate. This sediment will, as a result, not be deposited in areas where it once was; thus this can have knock-on effects.

Tidal flow

As well as governing the rate of water movement into and out of an estuary, tidal flow also controls the height of the tide. Modification of the flow can cause the tidal range to vary; that is, if flood tide flow is restricted by constraining it through a narrow opening, there is going to be a backing-up of water, and thus reduced tidal heights upstream of the structure, and increased tidal heights downstream. Similarly, on the ebb, the ponding effect will occur upstream of the barrage, prolonging the period during which intertidal habitats are submerged. If you vary the tidal elevation, you could alter the depositional regime of the area in one of two ways. First, marshes could become emergent – that is, no longer covered by the tide as frequently or at all – or secondly, marshes could become inundated more frequently, and thus might undergo species change, or even drowning.

Tidal flow also governs the physical form of the estuary, and the way in which river water (fresh) mixes with sea water (salt). Salt water is denser than fresh, and so unless these two bodies of water are physically mixed by currents, it is possible to have a situation where a salt wedge is formed; that is, where the incoming tide pushes a wedge of salt water under the fresh river water. This salt wedge moves in and out with the tide, and as it does so, it provides a powerful aid to the flushing mechanism of the system. If a barrage is built, this salt water may be impounded, and unable to

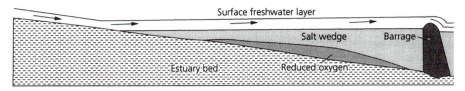

FIGURE 4.6 Simplified section of the impounded water behind the Tawe barrage, south Wales
Source: WWF-UK (Peter Dyrynda)

migrate. In the Tawe estuary in south Wales, a captive wedge of salt water is retained upstream of the barrage, and is only occasionally flushed out by extreme river flows. Periods of neap tides and low river flow produce periods of stagnation and reduced oxygen levels in the barraged lake (Figure 4.6).

Retention of sediment

Such a blockage to water flow as that mentioned above is going to cause many knock-on problems to occur. As well as large amounts of water moving on each tidal cycle, there are significant volumes of sediment which also get shifted around. A barrage holds back large quantities of sediment which become deposited on the bed of the estuary upstream of the barrage. This causes two problems. First, the water volume is reduced by the increased space needed for the sediment. On a power-generating barrage, this means less water to flow through the turbines. It also means that the barrage will, in effect, begin to silt up. In rivers with a high sediment flow, this may start to happen fairly rapidly. To a certain extent, this decrease in volume due to sedimentation is compensated for by an increase in water level. On extreme events, this could lead to overtopping of the barrage, but more commonly leads to the inland movement of the tidal limit on all the rivers draining into the estuary.

Second, before the barrage, this sediment was transported out to the coast. With the prevention of this, sedimentary landforms downstream from the barrage may become starved of sediment, and hence begin to erode. This will, therefore, have implications for defence and many uses of the coastline, along the lines which have been discussed, and which are covered in Chapter 6.

Changes in currents

Modification of current direction will also impact on mudflat morphology. Changes in currents mean alterations to the bed shear stresses, and thus a reduction in stress caused by reduced current velocities is going to further enhance deposition and erosion problems downstream of the barrage, with subsequent effects on navigation and benthic communities. Any localised increases in velocities around the new structure will also cause scour.

Impounding of water

If water is impounded, then it is going to stay at the same elevation for long periods of time. This can cause several problems:

1   Concentration of wave activity at a particular level will occur, which can enhance marsh and mudflat erosion at specific levels.
2   Increased erosion means increased water turbidity, owing to increased suspended sediment.
3   Prolonged coverage by water means that environments become anaerobic, or certainly less aerobic.
4   Fixed barrages may cause reduction in tidal range, and thus a higher low-water stand which leads to habitat modification.
5   Higher water levels mean modifications to the local water table and the saturation of farmland, and possible alterations to the base levels of local rivers.

Apart from the physical environmental effects, the ponding up of water also has significant impacts on water quality. The washing down and entrapment of decaying organic matter can cause periods of oxygen starvation, further enhanced by the poor exchange of water with fresh, oxygen-rich water from downstream. Considering that upstream urban areas continue to flush pollutants into the river, these accumulate in the sediments and water of the impounded lake. Algal blooms can start to occur, as well as increases in levels of metal pollutants in sediments. Reduction of water quality is known to impact directly on wildlife, as well as on human usage of the lake for amenity purposes.

Local increases in mean sea level

Seaward of the barriers, impedance of flood tide flow can elevate sea levels owing to the backing up of water. The increase can be of the order of 10 cm for a large estuary, which, although not a large number, can have serious implications for ecology and marshes.

Loss of flooding

The idea of barriers is that they prevent flooding. While this may be fine for human activity, the loss of periodic flood waters can have a serious impact on natural ecosystems. Loss of a periodic inundation by saline water can destroy brackish-water lakes or cause an alteration in the balance of brackish-water communities, causing a shift to terrestrial vegetation, while the loss of periodic sediment input reduces nutrient input to land.

Wildlife

The muddy shores of estuaries are important for a whole host of wildlife. Any change in morphology or habitat can have serious implications. The ponding up of water reduces the intertidal area, thus reducing the available habitat for many organisms, as well as the feeding ground for birds. Studies in the Tawe estuary (Dyrynda, 1996) have shown that since the barrage was constructed, the densities of key invertebrate species have fallen sharply, removing the staple diet of many bird species.

Estuaries are also important in the breeding cycle of many fish, providing nursery and feeding grounds for many coastal fish. Problems of water quality, oxygen levels, and the physical blockage of an estuary can lead to population fluctuations and subsequent knock-on effects on local food chains. In addition, the presence of a major barrier across an estuary along with consideration of the importance of estuaries as overwintering sites can mean that such structures can have an influence on migration routes.

◆　◆　◆

The impacts of barrage construction have been clearly demonstrated for the Oosterscheldt storm surge barrier on the Dutch coast. This barrier was completed in 1986 and resulted in a large reduction in the frequency of marsh inundation (de Leeuw et al., 1995), partly as a result of a 35 per cent reduction in tidal range (de Jong and de Jong, 1995). There was also a recorded drop in numbers of some infauna species (Seys et al., 1995), although other species increased in abundance, producing no net reduction in total biomass.

The reduction in frequency of inundation meant that flooding of the marsh surface did not occur during the growing season, reducing the distribution of seeds. In addition, reduced inundation also produced a reduction in rates of vertical accretion because less sediment was reaching the marsh surface. Another result of reduced inundation was that high marsh species could tolerate lower elevations, while some mid-marsh species were no longer being inundated frequently enough. Hence, high marsh plants colonised the mid-marsh area, resulting in the disappearance of mid-marsh flora.

Other changes caused by the reduced inundation are highlighted by de Jong and de Jong (1995). Soil properties changed and became more oxic, resulting in mineral growth. The drying out of sediment caused bank collapse due to weakening by desiccation cracks, while the reduced sediment strength led to lateral marsh retreat. Other impacts from less frequent inundation were that some infauna disappeared because the sediments were too dry, while others became more abundant (Seys et al., 1995). This shift in the ecosystem structure could also have implications for wildfowl which feed on the marshes.

◆　◆　◆

Leaving aside the tidal potential in estuaries, offshore the energy of waves can also have great potential to generate electricity. While the methods of generating wave-based electricity are generally less destructive of the environment – that is, they involve no large-scale engineering works – it is not true to say that there are no detrimental effects. By effectively removing the energy from waves, you are reducing the ability of those waves to work on the coastline; that is, to erode. However, waves are important mechanisms in controlling shoreline processes. The shape, form, and behaviour of the coast are largely dependent on waves, particularly along the open coast. Loss of wave activity at the coastline will cause decreased erosion along cliff lines, reducing the amount of sediment available to the system, thus leading to accelerated beach loss in many areas down-drift. Such events will also have associated financial implications for seaside towns and resorts in the form of beach loss due to sediment starvation (see discussion in Chapter 3).

In contrast, the same process can also be used for the benefit of some coasts which have severe erosion problems. We saw in Chapter 3 how offshore breakwaters are being used increasingly to reduce wave activity at the coast. It is potentially also possible to utilise this form of power generation as a coastal protection measure. On some coasts, where erosion is a problem, reducing wave activity is a desirable factor, and so in such situations this may be a practical proposition, and one that offers a useful 'by-product' at the same time.

Summary of Chapter 4

It becomes quite clear from the series of issues raised in this chapter that the human impact on the coastal and estuarine system is far-reaching. Before this chapter, I was stressing the importance of sediments from the point of view of coastal erosion and sediment budgets. In this chapter, I have introduced another aspect of sediments' role, in that they are also critical in the storage and remobilisation of pollutants.

This clearly puts our consideration into a new realm, because the presence of pollution is, by and large, something which cannot be seen. The discharge of pollution to estuaries or coasts is now regulated in most industrial countries, although exceptions do occur. What is needed is some knowledge of how much pollution is actually held within the muddy sediments of the world's estuaries and coasts, in order to predict the significant unmonitored discharges which occur when systems start to erode.

Aside from the pollution problem, there is another aspect of industrialisation which is also important. With the abundant water supplies, it is no great surprise that many power stations, especially nuclear ones, are located at the coast. Some slight pollution problems can occur, typically thermal, although in the case of Sella-field radionuclide emissions occur as well. Newer, 'environmentally friendly' forms of power generation potentially have much greater impact on estuaries because they can significantly alter the whole regime. In Chapter 3 we mentioned the possibility of the large-scale remodelling of estuaries as a means of management. This idea is dismissed at present because there are too many unknowns in this approach. Rather strangely, therefore, there are many plans for barrage construction, which will achieve just that effect. Virtually all the environmental components of a barraged estuary will change, which appears acceptable where the approach is finance led, but totally unacceptable, and a dangerous venture into the unknown, when the approach is research and science led. There appears to be a contradiction in ethics and methodology here in the world of coastal management.

This chapter has been rather diverse, and in places rather specific. It has been necessary to introduce key points about the types of pollutant within a system in order for the reader to link the problems discussed here to other, more specific texts on pollutant behaviour and types. It has been the intention of this chapter to provide the reader with an understanding of the following issues:

◆ the historical nature of pollutant storage in muddy sediments;
◆ the potential of muddy sediments to act as future pollution sources;
◆ the role of power generation in coastal problems, and the significance of nuclear power for long-term coastal protection;
◆ the hazards of estuarine barrages and their implications for the stability of the estuary.

Coastal development
for tourism

Introduction

So far we have considered the coast as a place where industry is located, population centres develop, and ports and harbours become established. To a large extent, these populations are fixed and are permanently based in these localities. However, in addition to these uses, the coast also has great appeal to a large transitory population who use the coast for leisure and recreation. As a result, affected parts of the coastline have undergone development dedicated to the recreational needs of the tourist. We have considered the problems associated with coastal development in other chapters, and so these need not be reiterated here. The form of development associated with tourism is really no different from that associated with urban or industrial development; the problems of high hinterland values and the need for coastal protection are the same. What does make tourist development different, however, is its intensity, and the number of people located in any given area. In other words, the number of people per unit area is generally greater in tourist areas than in non-tourist areas of similar size, and so as a result the impacts are more concentrated.

Such coastal tourist development is not a recent thing. The trend for coastal holidays and leisure time started when it became fashionable to take the sea air around the late eighteenth and early nineteenth centuries (see Goodhead and Johnson, 1996, for a detailed synopsis). As a result, resorts started to develop in many parts of Europe and, by the late twentieth century, along the Atlantic coast of the USA. Throughout the twentieth century, the development of resort areas has spread along the coasts of most developed and some developing countries. With increasing freedom of travel during the second half of the twentieth century, the development of coastal tourism has spread, concentrating large resort areas in climates of dependable sunshine. This increased ability to travel has meant that more areas the world over are available for the tourist. As the number of people wanting to go to a particular resort area increases, so more and more areas develop into larger resort towns, with increased development for the provision of amenities and hotels.

Developing coastal areas for tourism involves many of the processes associated with developing the coastline for industry, ports, or power stations. A coastal resort needs access to the sea, inevitably meaning some form of sea front promenade, which often comes in the form of a sea wall. But resorts do not just mean hotels and sea front bandstands. The whole idea of coastal tourism, from a coastal management point of view, covers a series of major issues, including:

1   need for access to the foreshore, possibly involving stairways down cliffs, or paths across sand dunes;
2   pressure on coastal habitats, such as trampling pressures on grasslands and dunes;

3   need for a stable beach with suitable sediment. Beaches which are unstable will need to replenish their sediment levels at intervals, otherwise the tourist population will move to other resorts;

4   the problems of multi-users and conflicts of interest. People who want to swim cannot do so in areas where others are boating, for example. Increasingly, holidays are becoming primarily active rather than passive. People want to do things, such as boating, water-skiing, snorkelling, etc., rather than lie on beaches;

5   increased pollution caused by the large influx of people. Pollution can take many forms, including sewage, litter and traffic fumes;

6   requirement for facilities to entertain the tourists, which can dramatically increase the size of resorts, resulting in sprawl. In addition, other facilities such as leisure parks, entertainments and marinas all increase demands on the local environment.

Many researchers regard the leisure industry as having the greatest impact on coasts and estuaries. With the expansion of leisure time and the increase in demand for facilities, the demands on estuaries and coasts are increasing, and with such an increased demand, there is an increased need for careful management in order to protect sites for future uses, and the coexistence of other interests. The importance of tourism cannot be understated. Miller (1993) highlights the significance of tourism from different viewpoints, and indicates that 12.3 per cent of the world's total consumer spending goes on tourism, creating jobs for 6.5 per cent of the world's workforce. Clearly, with such a world commitment to this one industry, there is great pressure to maintain tourist popularity, and, as a result of this, environmental concerns often come a poor second to commercial demands.

Some coasts experience greater impact than others. Tidally dominated coasts such as estuaries, for example, do not generally appeal for beach holidays, but more to bird watchers, walkers, wildfowlers, etc. Such people tend to avoid the kind of large-scale tourist developments which can mean serious environmental degradation, because the coast's natural state is essential for the attractions which they go there to enjoy.

Wave- and wind-dominated coasts, however, are much more popular, owing to the prevalence of coarser sediments in the form of sand and shingle beaches. Such features make suitable sunbathing sites, and so these areas are often heavily developed and heavily used. In addition, in marked contrast to tidal environments, because wildlife or natural history is not the prime interest, the preservation of natural habitat is not the prime concern; the priority is a good beach for people to lie on, and this can be achieved without the need to protect habitats. This means, therefore, that wave- and wind-dominated coasts have experienced some of the greatest impacts from tourism, and some of the greatest environmental degradation.

Because beaches can be viewed purely as an amenity – a means to attract the fee-paying visitors into a resort – they are often considered in isolation from the land. We have already seen that natural sediment movement needs the mobility of sediment along large stretches of coastline. To a resort, the main concern is to have a beach fronting it, and many resorts have their own set of defences protecting their own frontage. As we have already seen (Box 3.2, p. 77), such a management style does not facilitate good coastal management and leads to increased problems elsewhere.

As well as coming to lie on beaches, tourists continue to be drawn by coastal attractions such as cliff scenery and coral reefs, and so it is important that these areas maintain a degree of naturalness. In Australia, for example, reefs have become particularly susceptible to change from people going to look and to take souvenirs. The result is that visitors, albeit unintentionally, can damage the area which they come to see. Coral reefs are discussed more fully later in this chapter, with respect to tourism impacts.

Problems of tourist development at the coast

With so much pressure to cater for large numbers of people in relatively small areas, the competition for space in resorts is great, with the most being made of any that is available. Because of the need for good beaches, it is generally not policy to encroach on the foreshore area for development where beach holidays are the prime concern. It is common in areas of intense development pressure for adjacent estuaries to be seen not as wildfowl sites, but as potential marinas and sites for land claim in order to increase hotel space or leisure facilities. As a result, some resorts, especially in areas such as Florida and California, and Australia's Gold Coast, have dramatically increased their land area by claiming wetlands. Towle (1985) reports on a project at Gros Inlet in St Lucia in the Caribbean, in which wetlands were claimed for building land and for amenity development. Here, land claim was carried out in 1969 in part of the inlet to create dry land and an artificial lagoon to enable expansion of the adjacent resort at Rodney Bay. In this example, the loss of the wetlands was also considered an advantage because of a long-term problem with sand-flies, which had become a nuisance in the resort area. Although the land did allow the resort to expand, the area of land claim significantly modified ebb and flood tide currents, directing stronger currents onto the beach areas, resulting in increased erosion. In addition, local fisheries were affected – a negative economic impact – and the problem with the sand-flies remained. From this example, it can be seen that the environmental implications of interference remain, irrespective of the reasons for development (see Chapter 3). In the above example, the financially driven need for resort expansion has only served to cause a new problem of beach loss, which itself may have serious implications for the future prosperity of the resort.

In other tourist areas, the provision of marinas has also been a prime aid to attracting visitors. Small leisure craft need areas in which to dock, and so sheltered marinas are constructed, generally at great environmental cost to the local estuarine resource. In the UK for example, there is not one major estuary without a marina actually in place or planned (Figure 5.1; Pye and French, 1993a). In other resort areas the need for berths has been taken to the extreme, with the construction of facilities for the berthing of larger cruise ships. The construction of such a facility at Montego Bay on the northern coast of Jamaica has occurred in conjunction with resort development and the construction of artificial beaches which extend into the bay. It is possible to

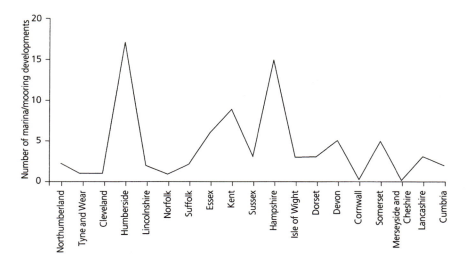

FIGURE 5.1 Number of proposed marina and mooring developments, by county, in England
Source: Based on data from Pye and French, 1993a

identify major problems with this form of development. Clearly, this area, Jamaica's major tourist centre, needs attractions and the facilities to attract tourists, and large amounts of money have been spent on their provision. Looking more closely, however, one begins to question the practicalities of such moves. For example, was it sensible to claim areas of mangroves (shallow-water tidal areas of high sediment deposition) in order to construct a deep-water berth for cruise ships? What has been the impact on the local tidal regime and sediment transport patterns? Second, when artificial beaches are being constructed, it is perhaps pertinent to ask why there are no beaches there already; perhaps the environmental conditions are not suited to sediment deposition. In such cases, more sediment can be brought in each year or so to recharge the area to maintain its attraction to tourists, but what happens to the sediment which is lost?

This example therefore clearly shows what may typically happen in some resort areas. Instead of development going along with the constraints and limitations of the coastal environment, as is recommended by theories of coastal zone management (see Chapter 7), the natural environment is being made to fit the development. Clearly, the construction of beaches where they do not naturally occur, and the construction of major deep-water areas where they are not found, are two prime ingredients of coastal instability.

The development of tourism on islands where land is scarce encourages coastal tourism. In Japan, there has been much coastal land claim in order to create open spaces and facility development. The Kasaioki scheme in Tokyo Bay was a major 380 ha land claim and development scheme which occurred during the late 1980s and early 1990s, and has allowed the development of the Kasai Seaside Park (Hudson, 1996) . The neighbouring Tokyo Disney World is also built on claimed wetlands at the mouth of the Edo river. In addition to this, many schemes around Tokyo Bay have claimed wetlands to provide beach areas, open spaces, and wildlife parks. Such significant areas of land claim will cause extensive modifications to the tidal and wave regime (see Chapter 3).

Impacts of tourism on dune coasts

Perhaps one of the greatest areas of tourist impact is in sand dunes. Dunes are very fragile habitats and, from a process point of view, represent the temporary store of beach sediment in the supratidal zone. Sediment which is regularly moved and reworked onto the beach tends to be highly mobile, while that which has been in the dune system for greater periods of time tends to become vegetated and partly stabilised. Although the presence of vegetation provides increased stability, the non-cohesive nature of the sediment causes it to remain quite mobile, and although it can withstand wind, the vegetation is rarely sufficient to withstand the impacts of trampling, off-road vehicles, horses, or trail bikes.

It has been shown both by field observation and by experiment that any area subjected to visitor pressure will, at some time, show signs of degradation. This degradation is manifest in several ways. One way is that the total biomass of plants is reduced, leading to a situation where species are lost and replaced by trample-resistant varieties. While in coastal heaths or grassland this species change may be possible, the range of vegetation which can tolerate the harsh regime of sand dunes is limited, and so large-scale species loss tends to happen rather rapidly. The total loss of species can also result if trampling increases sufficiently. Work done on the Winterton dunes in Norfolk, eastern England (Boorman and Fuller, 1977) clearly demonstrates a correlation between areas of intense trampling damage and pathways, with the more heavily used pathways showing greater degrees of trampling and floral degradation. Grasses such as marram and lyme are most easily removed by trampling, but are also the most effective at accumulating sand.

Once sand dunes have started to lose their vegetation cover, the sand becomes more mobile because it no longer has its protective cover of vegetation with its roots which bind the sediment together. This will allow the wind to start moving the sediment around, causing blow-outs. These blow-outs then make ideal sites for picnics or parties, and so their usage intensifies. In England and Wales, the greatest human impact on dunes is from human trampling (Ranwell and Boar, 1986). Part of the problem arises from the fact that most people want to be at the sea-shore, but to get there they have to gain access through the sand dunes, hence forming clear paths. It does not actually take much trampling for an effect on vegetation to begin. Work by Boorman (1976) found that around ten passes in a month were enough to reduce the height of grasses by 66 per cent, and that forty passes per month would reduce it by 75 per cent. Around eighty passes per month would start to produce bare ground, while 150 passes could lead to up to half the vegetation being lost.

The leisure use of dunes, however, is not restricted to the impact of human trampling. Tourist use involves several other aspects, each of which can lead to damage of this fragile environment. There is an increasing tendency for leisure time to be spent in off-road vehicles. Tyres can behave in the same way as human feet, in that they produce a trampling effect and eventual reduction in vegetation diversity and abundance. In addition, however, the increased weight of vehicles leads to greater soil compaction and reduced oxygen availability to roots. Another difference is that damage due to vehicles is generally not as widespread as that caused by human trampling because vehicles are restricted to flatter, more consolidated areas. Regardless of this, 200 vehicle passes are sufficient to reduce ground cover by 50 per cent in summer, and by more in winter (Liddle, 1973). The effect of tyre damage is also long-lasting, meaning that sand is exposed to wind and subject to erosion for long periods of time. At St Ouen's Bay, Jersey, tyre damage was still evident in the area five years after vehicles were banned. This tyre damage comprises not merely lack of vegetation but also the displacement of large volumes of sand. As a result, regularly driven tracks become considerably lowered, in some cases by up to $0.6\,\mathrm{m\,yr^{-1}}$ (Leatherman, 1979; Leatherman and Godfrey, 1979). These problems are further exemplified in a study by Anders and Leatherman (1987) at Fire Island, New York. They noted a significant change in vegetation after only twelve weeks of use, with a retreating vegetation front and increasing dune instability. Furthermore, such changes become evident at only low-level use by off-road vehicles. Heavier use caused the total elimination of vegetation after just one year. Investigations also revealed that cessation of use will allow some regeneration, but this is severely reduced in abundance and vigour, and rapidly disappears with resumed activity. McAtee and Drawe (1980) demonstrated similar problems of vegetation reduction and loss at North Padre Island, Texas. They found that the degradation of vegetation cover was directly related to the type and intensity of traffic.

Horse-riding can cause similar levels of damage to vegetation, with the amount of damage increasing with the speed of movement: galloping horses will cause more damage than trotting, which causes more damage than walking. Other causes of dune damage occur in areas where a specific activity attracts interest. Sand-skiing and sand-yachting are spectator sports, and the natural relief of dunes favours them as a form of temporary grandstand. This form of damage is very local and does not affect all dune systems, but with four thousand spectators recorded at a sand-yachting event in Caithness (Ranwell and Boar, 1986), and considering the amount of trampling which can cause damage to vegetation, local impacts can be severe.

Increasingly, the demand for more leisure amenities is also having an impact on dunes in an indirect way. There has recently been a great increase in the number of people playing golf, and coastal links courses are well favoured. There are, however, two sides to this argument. First, the conversion of dunes to a golf course serves to restrict public access, reduce trampling, provide positive management practices, and preserve natural habitats – all of which may be considered as beneficial to dune areas. In contrast, however, it also requires the improvement of parts of the vegetation cover. While natural dune vegetation is fine for areas of rough and fairways, it is not really suitable for greens, leading to the planting of alien species of grass. These species are not generally as resistant to the prevailing conditions. Flooding with sea water at a golf course in Kent in south-east England (Ranwell and Boar, 1986) resulted in severe damage to the alien turf, while natural grasses were undamaged. Such flooding may be considered a rare event, and while the planting of high-quality grasses on greens may introduce a degree of unnaturalness to the environment, the concept of converting areas of dunes to golf courses can provide a good method of conservation. Although the management as a golf course could hardly be considered to be managing the system as a sand dune system, there are many areas which do maintain their natural form, and have significant conservation importance. There is also the added factor that the use of sand dunes as golf courses is not necessarily irreversible. Provided the conversion practice does not alter the morphology of the dunes in any great way, then the system could be allowed to revert to natural dune vegetation. This is not a simple process, however, because while in use as a golf course, the dunes are not receiving the same degree of sediment input, or allowed to interact with the natural environment in the same way as they would if left alone. This problem of sediment mobility can also produce further problems. For as long as the dunes are being managed for some other primary purpose, they cannot serve their role in coastal defence, unless provision is made for this in the siting of tees and greens. It is important, from a defence point of view, that these areas of dunes can continue to interact with the shore and allow the interchange of sediment with the beach.

It is clear then that from a tourist and leisure perspective, sand dunes represent an environment which is subject to many forms of human intervention. As a result, there are many management practices which are relevant to dunes. When considering the management of heavily used dunes, it is important to bear in mind that as well as providing a good amenity area which needs to be preserved, dunes also serve two natural functions. First, they provide a temporary store of sediment to allow for short-term adjustment of the beach during storm conditions, and second, they provide effective defence for the hinterland. Consequently, the management and protection of dunes has to be carried out in such a way that they are protected while still maintaining public access. As part of this access problem, it should also be remembered that most people who walk over dunes do so only as a means of getting to the beach beyond. Because of this, the provision of routes through dunes is perhaps of prime importance. There are three approaches to dune management. First, in heavily degraded areas it may be necessary to consider the construction of artificial dunes. This is perhaps the most drastic in respect of coastal management because it has huge implications for the sediment budget owing to the large volumes of sediment required. It is certainly true that the supply of sand would be naturally available, but at the expense of other habitats down the coast. Despite this, however, artificial nourishment (see Chapter 3) has been carried out in many dune systems. Adriaanse and Choosen (1991) discuss the methodology behind this procedure, while Baye (1990) discusses the construction and ecology of dune reconstruction in Canada; Cortright (1987) looks at management of dunes at Nedonna Beach, Oregon, and Nordstrom (1988) looks at the artificial grading of dunes along the Oregon coast. Second, the restoration of degraded dunes can occur in response to the sorts of problem which we have indicated above. And third, we could adopt the policy that to prevent is better than to cure, and undertake a programme involving the management of healthy dunes.

When dunes are being managed, by whatever method and approach, it is important, from a coastal process point of view, to bear in mind the method by which sand is transferred to them. Sand is derived from the intertidal zone at low water via the wind by a process known as saltation (the bouncing of sand grains along the beach). As a result, any structure placed along the ground will intercept the sediment. It is also the case that the sediment transport rate is proportional to the cube of the wind speed, and therefore only slight variations in wind speed are needed to increase or decrease the amount of sand transported to the dunes. The critical speed required to initiate sand movement is variable according to the grain size concerned, but is generally in the region of $4\,\mathrm{m\,s^{-1}}$ (Bagnold, 1941).

The most common way of encouraging the accretion of sand is to impede the wind velocity, typically by installing 1 m-high wire mesh, brushwood, or geotextile fences, which interrupt the coarser sand but allow the finer material to pass through and on up into the higher dunes, hence ensuring that these areas do not become starved of sediment. The effect of these fences is felt approximately eight times the fence height downwind. In other words, a fence of height 1 m would allow up to 8 m of land behind the barrier to undergo some degree of sedimentation. Once the sand is deposited and has buried the fences, planting of vegetation can stabilise it.

Where dunes have been badly damaged and need to be rebuilt, re-forming them using earth-moving equipment is a method that can be adopted, although this can have implications for beach levels and sediment supply. Alternatively, the stabilisation of eroding dune faces with some form of mesh can be useful but tends to prevent any further process, such as transfer to the dunes inland. All the above involve the restoration of the sand dunes themselves. Were matters to be left at this, then continued use by people would allow the problems to recur. Therefore, it is also important to bring in a form of 'people management'. Bearing in mind that most people use the dunes only as a route through to the beach, one way of doing this is to channel people along wooden walkways (Plate 5.1). This is a useful way of reducing the effects of trampling, and, as can be seen from Plate 5.1, the vegetation on either side of the walkway remains healthy because there is a great tendency to concentrate walking on these pathways, since the sand itself is very difficult to walk through. Other techniques can include excluding visitors from certain areas of the dunes by fences and keeping out grazing livestock. The problem here is that the wooden fences which tend to be used for this purpose also make very good firewood for dune parties, and accordingly many get damaged. Better success has been observed in areas where these forms of dune management have been coupled with a programme of public education in the form of information boards and displays. It appears that if the public can see why some course of action is being taken, then they are greater respecters of it.

Because dunes serve two important functions (as tourist amenities and coastal defence), their management is seen as critical in many countries. In the Netherlands, where much of the coast is claimed and around 33 per cent of the country is below sea level (Louisse and van der Meulen, 1991), maintenance is critical. Louisse and van der Meulen (1991) and Verhagen (1990) both identify the essential nature and importance of dune management along this coast. The Dutch government has spent

PLATE 5.1   A simple trample prevention management measure in sand dunes
(Ainsdale Coast, NW England)

a good deal of money investigating solutions to the management problems, concluding that the dunes should remain as natural as possible, although direct interference is worthwhile in areas of high hinterland value. Where possible, this will be in the form of soft engineering methods.

In Brittany, coastal tourism has been a threat to dunes since the Second World War (Guilcher and Hallégouët, 1991). Management has occurred without restricting public access to the beach, thus maintaining the tourist potential of the area. The planting of marram and the construction of sand fences and mesh matting have been successful in regenerating the dunes by natural processes of sediment entrapment and vegetation colonisation.

In the Mediterranean, where tourist pressure can be particularly high, management of dune systems is also of great importance to maintain tourist interest in the area (van der Meulen and Salman, 1996). Dunes here have been stabilised by many techniques, including renourishment, sediment entrapment, and afforestation.

Environmental impacts of dune management

The successful management of dunes is beneficial in a number of environmental ways. First, it provides important coastal habitat which acts as an effective line of soft defences. Second, healthy dunes provide a temporary store of beach sand which can be used during storm periods to supply the beach, as under storm conditions waves may reach the dune areas. Waves also remove much material from the beach by drawdown, hence, by using the dune sand, the beach sand can be replenished, and beach level maintained. In non-stormy conditions, the sea will bring this sand back in to the beach, where it will be returned, via the wind, to the dunes. The management of dunes can have significant environmental benefits for coastal defence and beach stability provided that it occurs in a way which maintains the ability for land–sea interactions.

Overmanagement is always a problem in dunes. If the dune becomes over-stabilised, then its store of sediment will become locked up in the dunes and become unavailable to beach processes. The result will be a coast which cannot adapt in order to absorb wave energy efficiently, which will have implications for long-term beach stability. This overmanagement can result from the overuse of fencing and other sediment entrapment methods which may be fine in trapping the sediment to build up the dunes, but will not allow the movement of sediment seawards again. The effect is the same as that which happens when a sea wall is built along the face of a cliff: a supply of sediment to the coastal sediment budget is removed.

Impacts of tourism on other coastal habitats

While sand dunes represent perhaps the area of greatest environmental impact caused by tourism, other coastal habitats also suffer their own forms of environmental impacts. On salt marshes, the constant use of the upper marsh by walkers can lead to problems of trampling and the loss of vegetation. Mudflats and sandflats rarely suffer damage, partly as a result of their inaccessibility, the danger of walking on them, and the ability of the tide to repair any damage which may be caused. Mangroves are environments similar to salt marshes but are even less accessible, experiencing few direct impacts of tourism but many indirect ones, such as land claim and conversion to marinas, leading to problems of sediment storage, hydraulic changes, and pollution.

Coastal habitats found in wave-dominated environments do tend to experience greater tourist and leisure impacts. Many resort areas owe their existence to beaches, whether shingle or sand-dominated, and so these areas tend to be the aspect which tourist authorities need to have right. Many of the methods used, however, can contravene the best practices for coastal management. Natural processes build up beach profiles and move sediment along the coast, both processes which are important to the overall management of the coastal cell. Sediment loss due to longshore drift, or to defence works up-drift, can have serious implications for resort areas as they see their beaches disappear. As a result, areas of coastline where resort areas proliferate also tend to be some of the worse cases of piecemeal and poorly judged sea defence construction. Many local authorities, in response to requests from their tourist authorities, construct defences to protect their beach sediments, resulting in there being no overall management structure for these coasts. Another approach is to feed the beach. Being a soft engineering method, this is more acceptable from a coastal management point of view as it allows the continued movement of sediment along the coast. In addition, however, by feeding the beach with sediment, the area can also be maintained aesthetically. Bournemouth, on the south coast of England, is a major resort area. The loss of sand from its beaches became a serious issue in the late 1960s and early 1970s, with the result that a major recharge scheme was undertaken in 1974/75. Continued longshore movement of sediment meant that this had to be repeated between 1988 and 1990, with a further scheme expected around 2003 (Cooper, 1996). By adopting this method of foreshore generation, the Bournemouth authorities have achieved two major objectives: first, they have secured a nice sandy beach to support their tourist industry; and second, from a coastal management point of view they have done this in a way that does not starve down-drift areas of sediment. There are many examples of beach renourishment, some in response to tourism and some as a means of compensating for low beach levels. A select bibliography can be found in Table 3.4 (p. 87).

Impacts of tourism on coral reefs

Second to sand dunes, coral reefs are perhaps the most tourist-damaged habitats found on the coast. Being located in warm, shallow clear waters, and containing a very diverse range of fauna, reefs have an immediate appeal to the tourist. Damage to reef sites by tourists can be quite serious, often purely as a result of lack of knowledge of the implications rather than by any deliberate act. Collecting of specimens and trampling are two of the major historical causes of reef destruction, with additional problems caused by the berthing of boats. While much damage has been done in this way, these are all problems which have been realised and tackled by the setting up of conservation parks and management schemes.

In Sri Lanka, tourism, along with other land uses, has resulted in a steady depletion in the extent of reefs (Rajasuriya *et al.*, 1995). Rapid and largely unplanned tourist expansion in the late 1980s and early 1990s has had severe impacts on the reefs, partly directly, but mostly indirectly via increased sewage and sediment loads due to land clearance. In this example, much of the change has occurred in areas of conservation designation, such as in the Hikkaduwe Marine Sanctuary. Collection of specimens has been banned since 1929, and many sanctuary areas have been declared since then. This has caused a great reduction in the direct impacts, although the indirect problems have not been tackled, meaning that the resource is still threatened. In the Philippines, reefs are said to be in a reasonable condition (Gomez *et al.*, 1994), although they still suffer from the problems mentioned above, again owing to tourist development. In the Pacific, reef health varies directly with the intensity of development. Areas distant from any tourist or population centres are healthy, while those closer to such areas are degraded (Maragos, 1993; Zann, 1994). In French Polynesia (Hutchings *et al.*, 1994), similar degradation problems occur, as they do in the Galápagos Islands (Glynn, 1994) and Thailand (Hale and Olsen, 1993).

It appears from all this evidence that while management clearly plays an important role in reef protection, it appears to solve only part of the problem, because it covers direct impacts such as those of boats, fishing, collecting, etc. Management tends to overlook the indirect problems, such as development, sewage, sediment loads, etc., which, from our examples above, appear to be the main causes of reef degradation. The problems of increased tourism and related increases in sewage disposal, poorer water quality, and sedimentation clearly cause impacts on a habitat which depends for survival on clean water.

Kanehoe Bay in Hawaii is an area where reefs and tourism have to coexist. Better access and links to major road networks opened up this area to settlers as well as tourists and, as a result of the subsequent increased sewage discharge into the bay, coral populations started to decline and large sections of the reef became impoverished. Because of increasing concern for the coral environment, all sewage outfalls were redirected away from the bay by the 1980s, with the result that by 1983 nutrient levels had declined, with corresponding increases in water clarity and quality. Shortly after this, corals started to regenerate (Jokiel *et al.*, 1993; Maragos *et al.*, 1985; Smith *et al.*, 1981). This example clearly shows how rapidly reefs can become damaged, but also how robust they are and how rapidly they can begin to repair themselves. Guzmán (1991) details another example of reef restoration in Costa Rica. In some areas, all coral had disappeared as a result of human impacts and natural processes. While this may not be entirely attributable to tourism, the restoration method does provide a good example of how tourist (and other) impacts can be rectified. Around Cano Island, coral fragments were imported from a nearby reef and transplanted onto dead reef structures. After three years, around 83 per cent of these were still alive, and some had naturally split and colonised additional areas. Clearly, transplanting coral may provide a method of reducing the degradation problem, but this can only occur once the original cause of degradation has been removed; that is, sewage has been stopped, water clarity has increased, or collection has been banned.

In other cases, the original degradation problem has been caused by the local population, as has happened in Sri Lanka, where around 40 per cent of raw material for the cement industry is derived from corals (Lindén, 1990), and in Madagascar, where whole corals are utilised for building. Apart from building, the fishing of reefs can also lead to problems, especially when the techniques employed involve the use of dynamite, because while killing the fish, which then conveniently float to the surface, one stick of dynamite can also destroy all organisms over an area of up to $100\,m^2$ (Lindén, 1990). For the sake of completion, it should also be noted that reefs are additionally subject to impact from rising sea levels. This issue is discussed in more detail in Chapter 7.

Demands of the tourist industry

So far we have looked at tourism-associated problems in specific coastal settings. In these cases, the problems discussed are somewhat location specific. There are, however, a series of problems which are common to a whole range of coastal environments.

Need for access

There is little point in having a tourist beach if people cannot reach it. We have already seen how sand dunes can be severely damaged by being used for access to beaches. The greatest impact of tourist access is twofold. First, if access is restricted by defence works, then many people are going to be concentrated into areas where access is possible. As has already been seen, this can lead to problems of vegetation loss through trampling, but can also cause excessive sediment compaction. Particularly in sandy environments, the compression of sand can produce areas of sediment which are firmer, and therefore more resistant to wave action. This can result in reduced infiltration of water into the sand and the resulting increased surface flow. On popular tourist beaches this effect can be significant, with increased surface flow causing increased erosion of the beach. The second problem with access is when it allows greater usage of areas which were previously isolated. Access to beaches through boats, for example, can cause disturbance to birds; similarly, many shingle spits provide homes to important breeding colonies of several species.

When it comes to coastal management, it is clearly necessary to build access into any defence constructions which may occur. Similarly, access routes need to be built into any soft engineering structures, such as boardwalks through dunes, or 'bridges' across otherwise problematical routes to the beach (Plate 5.2).

Another aspect of access is that of the car. People who visit resorts, whether resident or day trippers, tend to arrive increasingly by car. The sheer number of

PLATE 5.2   Wooden bridge allowing beach access across large sea defence blocks
(Worthing, Sussex, UK)

vehicles can be problematical for many resort authorities, who may have to turn areas of flat ground into car parks. This land use can make further demands on already stretched land resources.

Pressure on coastal habitats

Large numbers of people and wildlife habitats rarely mix successfully unless careful management is practised. We have seen how tourist resorts can make great demands on intertidal sites for land claim and marina development. This represents not just a gain in land on which to build new hotels, but also a significant loss of intertidal habitat for wildfowl, and of sediment deposits which could act as a form of coastal defence. It is true to say that in many coastal resort areas, natural habitat is practically non-existent, and the whole coastal system becomes artificial. Once it becomes artificial, it needs constant intervention to maintain it in this form.

Pressures in tidal environments are perhaps limited, but on open coasts the demand for hotels and amenities tends to mean that resorts develop and encroach further and further into sand dunes, heathland, or lagoon areas. As an example, along the Sefton coast in north-west England, hotel development spread northwards along the coast from the Mersey estuary, including onto dune areas. What was particularly extreme, and perhaps rather unbelievable with hindsight, is that the demand to spread was greater than the suitability of the land, with the result that hotels were built with views of the back of the sand dunes, rather than the sea. As a result, the tops of the dunes were actually removed to provide a sea view to guests of the hotels. Clearly, such demands on coastal habitats are well beyond the recommendations of any coastal management plan and only serve to reflect the misconceptions that habitats should come second to development.

Need for a stable beach

No coastal resort can survive without its beach. Tourist authorities will go to great lengths to protect their prime assets, whether it is by feeding or by means of hard structures. We have already seen, both in this chapter and in Chapter 3, what can happen to the coast as a whole when too many authorities start to interfere with little bits of it. To a certain extent one has to be sympathetic to hotel owners in towns where coast erosion removes their beach. Changes in coastal deposition and accretion patterns do happen, but, while a new beach may start to accrete elsewhere, the resort is fixed to its location. As a result, coastal management has to allow for such situations and has to recognise the fact that in some circumstances it is necessary to force a depositional environment to stay where it is.

Another aspect of this problem is where the depositional environment changes from being sand dominated to mud dominated, or where sandy beaches are invaded by unwanted vegetation. There have been many examples of *Spartina*, a salt marsh coloniser species, invading beach areas. Southport, in north-west England, experienced this problem in the 1970s and 1980s when vegetation spread from the adjacent Ribble estuary. In this case the vegetation was removed to maintain the beach as a sand-dominated feature. Further problems occur when the environment shifts from sand to mud. At Grange-over-Sands, at the mouth of the Kent estuary in Morecambe Bay, the sandy beach has been replaced by mud, with the result that salt marsh has now developed. Clearly, this has had an impact on that particular aspect of the town's tourist trade, but has occurred as a result of natural processes within the channel. To date, the exact causes of the change are not known, but the natural shifting of the estuary's ebb channel away from one side of the estuary to the other, bridge construction upstream, defence works, and a newly installed breakwater may all have some responsibility.

Problems of multi-users and conflicts of interest

One of the greatest problems of leisure and tourism is that often it is a case either of too many people wanting to do too many different activities in too small an area, or people wanting to carry out certain activities in areas which are not suited to that activity. Perhaps one of the greatest areas of conflict is that which exists between leisure pursuits and nature conservation. Such conflicts can be categorised (Sidaway, 1988; Sidaway and van der Voet, 1993) as disturbances to birds or other animals by noise and contact, especially at certain times of the year. Such problems can lead to the loss of young to predators due to nest abandonment, or to displacement of breeding colonies. Another conflict can arise as a result of activity and damage to soils and vegetation. We have already discussed trampling and its effects in detail, but severe trampling can also cause loss of nest sites and loss of foraging habitat. Specimen collection, such as collection of eggs or individuals, can also cause reductions in biodiversity. This problem can also be extended to cover angling, where local reductions in species may occur.

Apart from nature conservation, the conflicts of use which occur along the coast often increase pressure for more of the coastline to be turned over to leisure pursuits, and to undergo related tourist development. The obvious resources for which people come to the coast are the water and beaches. People lying on a beach may find shore-fishing, power-boating, dog-walking, or jet-skiing incompatible. Similarly, swimmers may find the presence of boats or jet skis a hazard. This leaves a zonation problem in many resorts, although there is no reason why all activities should not occur at the coast, and even within the same resort. A launching ramp could centre as an area of boating, where swimmers are banned. Similarly, swimmers and sunbathers could occupy the same part of the coast, as the two activities are compatible. The whole issue of usage zonation is one which is becoming increasingly used in coastal management as the numbers of people using the coast, and the number of activities in which they are involved, increase (see Chapter 7). It is seen as a good way of removing conflict, and also to reduce heavy impacts on particular parts of the coast. For example, the construction of a launching ramp could be located a short way outside a resort, so as to allow the launching of boats in relative safety away from bathers. Also, this could spread the intensity of vehicles. The proviso, of course, is that such a development should not infringe greatly upon natural habitats.

Increased pollution

Wherever large numbers of people gather, there is bound to be an increase in the amount of pollution. On beaches, this may be in the form of litter, which tends to have more of an aesthetic impact than anything seriously toxic, but can result in the mortality of shore birds: fishing line and plastic drinks-can holders can tangle a bird's legs, for example. Litter is an increasing problem in many coastal tourist areas (as well as many others). Clearly, a heavily littered beach is not appealing to visitors, although it is the visitors who cause the original problem. Coastal authorities are often landed with a cleaning bill in order to keep visitors coming to their resort. In South Africa, around R 8 million was spent in 1994 clearing beaches of litter (Ryan and Swanepoel, 1996). Not all of this litter is tourist related, however; a large quantity is derived from offshore. Along the Mediterranean coast, a similar problem was identified by Gabrielides *et al.* (1991). This study showed a direct relationship between the amount of litter on a beach and intensity of usage.

Another aspect of pollution, which holiday-makers do not always see, concerns water quality. The large increases in population which result from tourism ultimately lead to increased quantities of sewage and waste water in general (as we have seen from its impact on coral reefs). In some parts of the world, sewage is still discharged untreated, although increasingly treatment is occurring as a means of improving coastal water quality. Significant in this in Europe is the European Bathing Water Objective (76/160/EEC) with proposed revision (COM 94/36), which has committed England and Wales alone to spend £2 billion on improvements to water treatment (Drury, 1995). This directive will implement improved emission standards around the European coastline.

General water quality is reflected in several ways. Highly enriched areas may suffer from algal blooms and other forms of nitrification and eutrophication (see also Chapter 6), which represent both an aesthetic problem and a potential toxicity problem for marine life. In other ways, illness as a result of swimming is also a real problem in many resort areas.

◆　◆　◆

We have seen many ways in which tourism can impact on the coastline. Some of these impacts relate as much to other coastal uses as they do tourism, and so it is questionable how much is directly attributable to each cause. It is perhaps important at this stage to place the issues discussed into their correct context. It is clear that at the local level, many impacts can be severe. On national and regional levels, however, threats to the coast are greater from pollution and development in general, and so the cause lies with human activity in general rather than solely with tourism (Sidaway, 1995). We can thus regard tourism as contributing to a problem which already exists in many cases, rather than being the sole cause of a problem. There are, however, some qualifications to this. Large tourist areas and resort complexes can cause widespread impacts due to land loss, land claim and interference with coastal processes in areas where no other development occurs. Similarly, the large-scale threat to sand dune environments is predominantly from the tourism and leisure sector, and so when a management strategy is sought here, it is this aspect of usage to which attention must be turned.

Examples of tourist impact on coastlines

The range of problems discussed above can be exemplified with reference to the following examples.

The Mediterranean

Each year, an estimated 100 million tourists visit the Mediterranean coast, making it the European centre for international tourism, receiving around a third of all international tourists. This has had a significant impact with respect to environmental quality, both directly because of visitor pressure, and also because of the increase in pollutants and wastes that need to be disposed of. Very little of the natural coastline remains within the popular tourist countries such as Spain, France, and Italy, a fact that has been caused by the wide-scale conversion of natural habitats, such as wetlands, forests, and maquis (Hinrichsen, 1990). The Catalonian coast of Spain is heavily developed for much of its 580 km, and 75 per cent of the Romagna coast of Italy is also now developed. In the 1970s, for example (Viles and Spencer, 1995), the south coast of Attica, Greece, between Athens and Cape Sounion was a combination of olive groves and fields, while today it is a series of tourist complexes and luxury holiday villas with litter, household refuse, and discarded builder's material being common in the area. Meinesz *et al.* (1991) highlight this problem for the Mediterranean coast of France. Developments covering 106 km of the 656 km shoreline have had severe impacts on intertidal and subtidal communities, causing high mortality among some species.

Much of the Mediterranean tourism is based around beaches, and there have been large efforts made to preserve and even create these habitats. Such projects have significantly altered the sediment supply off the coasts, leading to increased erosion in some places and increased sedimentation in others. There has been a tendency in some areas to develop resorts, and then import the sediment to make the beach. However, this has overlooked one major factor: if a beach does not have any sediment on it (i.e., it is a shore platform), there is a good reason for this. If you add sediment to build up a beach, then this sediment will go the same way as the original. In other words, by adding sediment to an area from which it will be naturally removed, you are inputting sediment to a coastal system, leading to deposition elsewhere, with possible problems impinging on ports or harbours, or the swamping of habitats with sediment. Anthony (1994) discusses this tendency towards artificial beach formation and indicates that 75–85 per cent of artificial beaches in the Mediterranean have had negative impacts on natural beaches. These impacts have been associated with erosion of the feed material, the deposition of this material on natural beaches (the feed material is coarser than the natural sediment, therefore once deposited on natural beaches, it becomes less mobile, and leads to artificial beach profiles), and beach shape.

In addition to beach erosion and modification, and loss of the coastal strip to development, there is also a serious pollution problem. We have already highlighted the litter problem which the area experiences (Gabrielides *et al.*, 1991) but in addition, the Mediterranean suffers from chronic oil pollution, most of which originates from the cleaning of tanks and discharge of ballast from tankers (Clark, 1989). Domestic waste, particularly sewage, is also a serious problem, especially around the major tourist centres. The problem of sewage is made worse by the fact that only around 20 per cent is treated before discharge (Buckley, 1992). While along many coasts the natural movements of water and the waves will break up and dilute inputs, the Mediterranean is generally non-tidal, the sea is shallow, and flushing into the Atlantic Ocean through the Straits of Gibraltar is a very slow process, and hence pollutants tend to accumulate in the natural environment (see Chapter 4).

The destruction of the natural coastal habitat of the Mediterranean has placed many species on the endangered list as their habitats are destroyed, and the disturbance from holiday-makers has driven birds away from nest sites, and destroyed feeding areas. Many of the coastal wetlands have been reclaimed and converted to other uses, largely associated with the tourist industry. In other areas, low-lying parts of the coast have experienced accelerated subsidence due to increased groundwater extraction necessitated by the large demands made by the seasonally high populations. This subsidence is also coupled with the increased saline incursion into fresh water supplies. Other reasons for subsidence reflect the fact that large areas have been claimed from wetlands. As we saw in Chapter 3, after being claimed, sediments undergo a period of dewatering and sediment compaction. This problem is particularly acute in the region of Venice, where a combination of land claim and water extraction has resulted in significant drops in land level.

It is clear that the Mediterranean has undergone radical change due to tourism, reflected in loss of coastal habitats, increased pollution, and loss of species. It is true to say that along the northern shore, where most of this intense development has occurred, the term 'natural coastline' has no place. The picture painted so far sounds bleak, but in spite of these impacts, there are many areas of attractive sandy beaches where people like to spend their holidays. The majority of these tourists, spending only one or two weeks a year on their well-earned holiday, will have no comprehension of the damage that has been inflicted on this environment.

There are clearly many management issues which need to be tackled. Pollution can be tackled at source, with increased water and sewage treatment and even tighter checks on tankers. The problem of coastal erosion is much more difficult because there has been so much human interference that any chance of re-creating a natural coast which can move sediment around and look after itself (see Chapter 3) has long gone.

The best hope for the area is a long-term strategy of regional coastal management. However, the problem with this is that owing to the intense tourist developments, any freeing up of sediment is going to mean the loss of some resort's beach. Similarly, the concept of allowing land to erode is also largely out of the question because there is little land of suitably low value.

It is clear that the northern (and parts of the southern) coast of the Mediterranean represent much of what is environmentally bad about coastal tourism. The creation of the mass beach holiday, while attractive to many people, has destroyed a large stretch of coastline. It is unlikely that the northern coast can ever be sustainably managed. Because beaches are so vital, sand continues to be brought in from the Sahara or other sources to feed beaches and make sure there is sand present for next year's visitors.

The Caribbean

As with the Mediterranean, some 100 million tourists visit the Caribbean each year. Although not as advanced, problems similar to those of the Mediterranean are threatening to overtake some of the islands, but because mass tourism has not yet had the same development impacts, and because of the morphology of the coastline (i.e. it is not practically landlocked), the problems of pollutants have not become as acute. Coastal development, however, is spreading unchecked, with even fewer planning controls than in Europe, causing large areas of natural habitats to be lost, and also adding unnatural constraints on the coastline. Tourism is acknowledged as the future of the Caribbean in order to support the population and to provide economic prosperity. At the same time, however, the local authorities realise that they need to monitor carefully the impacts of this growth on the environment. For example, disposal of solid waste, such as sewage, is a serious problem, with local seas becoming more polluted, with the result that coral reefs are dying. The unchecked development has also produced its own set of problems. On many islands, hotels have been built too close to the beaches, and because the sewage pollution is killing the reefs, problems of increased wave attack are causing increased beach erosion, which will eventually threaten to impinge on the protection of the hotels. The loss of beaches will become a serious issue as these are one of the prime ways of attracting tourists.

Much of the development has progressed at the expense of mangroves and forests, and supporting the tourist industry has caused overfishing and overcollecting of specimens from many of the reef areas. These are two main areas of concern. Another area is coastal pollution, as mentioned above. Less than 10 per cent of sewage and waste water emission from 170 million residents and around 100 million tourists is treated (Hinrichsen, 1990). Many of the urbanised coasts experience coastal waters full of raw sewage and litter, with resulting outbreaks of cholera and typhoid. This pollution problem, made worse by industrial pollution and agricultural run-off, is also leading to damage to coastal habitats, with damage recorded in mangroves, reefs, and seagrass, as well as to fish and sea-bird populations.

The loss of mangroves and coral reefs, coupled with the high pollution levels, also has one other major impact. Loss of reefs allows increased wave attack at the coast. Loss of mangrove means that sediment becomes less stable and more prone to erosion. Both these processes are driven by marine pollution. The combination of increased wave attack and decreased sediment stability means that coastal erosion will increase, and it has become a problem in Puerto Rico, Jamaica, Trinidad, and several US states on the Gulf of Mexico, such as Florida, Mississippi, and Louisiana (Hinrichsen, 1990). In order to combat this problem, stabilisation works are being carried out and mangroves replanted. However, this is rather piecemeal and clearly needs to be structured.

◆　◆　◆

The two examples represent two extreme cases of tourism impacts on coastlines. Nevertheless, the symptoms can be found, albeit to a lesser extent, on many tourist coasts around the world. For example, Hickman and Cocklin (1992) report on tourism and recreation development in the coastal zone of North Island, New Zealand; Yapp (1986) discusses recreation and tourism on Australia's coast; Charlier and de Meyer (1992) illustrate the threats posed by tourism along the Belgian coast; Smith (1992) highlights growing problems relating to the nature of resort development in Malaysia; and Carlson and Godfrey (1989) look at ways of managing the human impacts on a largely natural coastline in Massachusetts.

We have seen that different resorts cope in different ways with the range of problems which are created. Some realise what is happening early and start working with natural processes to provide a relatively cheap solution but one which takes time to achieve. Others go for the quick response; that is, if a beach is being eroded, they import a new one, only to have to repeat the exercise after a period of a few years. Such actions merely serve to keep the coast in an artificial condition, and do nothing for the overall stability of the regional coastline, or the advancement of coastal management.

Summary of Chapter 5

This chapter has, to a certain extent, drawn on information from other chapters and brought it together under the heading of tourism. It may appear from what has been said that all tourist activity at the coast is bad and is totally contrary to coastal management. While it is not true to say that tourism necessarily does much good to the coast, it is not much worse than many other human activities which we have discussed. We can regard tourism as one of those activities which have a bona fide case for being located in the coastal area, and so have to be accommodated in coastal planning. It can never be said that coastal tourism should be stopped because of its detrimental impacts, but it can be said that coastal tourism could be better managed to reduce some of these impacts.

We can, however, approach this in a slightly different way. If coastal tourism must, by its very nature, cause problems for any part of the coast on which it is located, would it not be possible to nominate parts of the coast in which to concentrate tourism? In other words, parts of the coast (possibly one coastal cell) could be sacrificed in order to protect others from any form of tourist development. Clearly, such suggestions are difficult in areas where a tourist industry is already developed, but in developing countries, where coastal tourism is still in its infancy, it may be possible to construct a more sustainable tourism strategy along these lines.

There are many links to be made between this chapter and others. The ideas of coastal developments and beach protection link to the discussion in Chapter 3. The ideas of pollutants are discussed in Chapter 4, and also partly in Chapter 6. On finishing this chapter, however, the reader should have an understanding of the role that tourism plays in coastal problems, in particular:

◆   the ways in which tourism can affect coastal areas;
◆   the problems associated with conflicts of use and the problems of relating conservation and activities;
◆   an insight into two actual examples of coastal problems, with references to others;
◆   an understanding of how this topic fits into the context of coastal processes, as discussed in other parts of the book.

Chapter 6

Causes of indirect impacts on the coast

Introduction

In the preceding chapters, we have considered the various ways in which humans have impacted on coasts and estuaries by actually interfering directly in coastal or estuarine processes, or by development of the hinterland. While such actions account for many of the problems which we observe around the world's coastline, there are other impacts which can occur many kilometres distant from the coast, but which have an impact on it. We can consider these impacts under two categories. First, there are impacts which happen inland from the coast, such as modifications to river catchments which can alter coastal configuration due to changes in the amount of water entering the sea, input significant amounts of pollution (see Chapter 4), or vary the volume of sediment. Second, there are impacts from offshore which are largely concerned with the pollution aspects of the coastline, and in particular the oil and gas industry and the subsequent shipping of oil-based products. Accidents to tankers can cause dramatic impacts along coasts, but actions such as the washing of tanks and dumping of ballast can also contaminate coastlines, and account for the greatest proportion of oil pollution. In addition to the problems caused by shipping, there is also the increasing problem of the disposal of old oil and gas platforms and, perhaps of more concern, of storage installations, such as the Brent Spar, which do pose huge potential threats to the coastline (see Box 6.1, pp. 182–3). Other offshore causes include dredging for sediments, which can alter benthic topography and, hence, the degree of wave exposure experienced by the coast.

Inland causes

While coastal management is an adequate tool for protecting the coast from activities which occur within the coastal zone, it also has to include other aspects which supply material to the coast, notably rivers. On some coastlines, a good deal of sediment which helps to build up the coast is derived from the river catchment, not from longshore drift, and so this sediment can be seen as being vital to the stability of the shoreline. Thus any activities in the river catchment which cause variation in the amount of sediment coming downstream can affect the stability of beaches in the same way that defending a cliff or erecting groynes can cause beach starvation.

Human impacts as a result of inland activities comprise those which occur within the river catchment. When soil is vegetated, it is considerably more stable than when left bare. The effects of deforestation, therefore, can have dramatic results for the amount of sediment washed into rivers. Miller (1994) estimates the amount of soil washed into lakes, rivers and the oceans each year to be sufficient to fill a freight train large enough to encircle the Earth 150 times. Deforestation also increases the amount of water entering the hydrological cycle because less is taken up by vegetation.

The way in which water is managed in a catchment can also be significant. Any modifications to the flow regime can impact on a coast if that particular part of the coast is in fine equilibrium between fluvial inputs and littoral processes. As a result, river engineering, flood protection, and channel works, all methods of river management with counterparts on the coast, can cause fluvial inputs to the coast to change, and, by implication, the coast itself to change.

Land use modification and the storm hydrograph

The storm hydrograph is a method of recording the flow of water in a river following rainfall. The form of the graph is a function of many environmental factors, such as intensity of rainfall, bedrock, soil type, soil cover, and land use, all of which help control the rate at which water is transferred from the land surface to the river network. Human impacts can have major influences on some of these variables, and so can cause changes to the hydrograph. Figure 6.1 shows two typical forms of a flood hydrograph. In Figure 6.1a, the discharge of a storm event is spread over a long period of time and tends to occur in a situation where water reaches the river by a combination of the various types of throughflow and surface flow. This form of the hydrograph is typical of rural areas with vegetation cover. In contrast, Figure 6.1b shows a sudden and short-duration peak which reaches the river rapidly, causing a rapid rise in water levels for a short period of time, then a rapid decline to previous conditions. This form of hydrograph results from a land use where the bulk of rainfall reaches the river via surface flow as a result of land modification. The most common way to have a system with mostly surface flow is where there is little soil and large areas of bedrock, into which water is slow to penetrate. Concrete acts in the same way as bedrock for hydrological purposes, and so the change of land use from rural to urban dramatically decreases the amount of water which can no longer penetrate into the soil. As development of the catchment continues and concrete replaces vegetation and soil, so the hydrograph can become more peaked, changing in form from Figure 6.1a to 6.1b. The area of Kanehoe Bay (Hawaiian Islands) has been directly affected by such processes. River catchments underwent large-scale development in the mid-part of the twentieth century (Jokiel *et al.*, 1993), which produced a large increase in

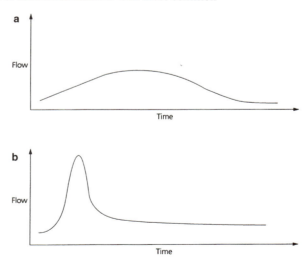

FIGURE 6.1 Hypothetical storm hydrographs: (a) typical rural catchment; (b) typical urbanised catchment

the area of impermeable surfaces. The subsequent change to the storm hydrograph and increases in sediment loads due to increased bank erosion blanketed the bay corals in layers of terrestrially derived sediment. While the direct impact was to kill the corals, secondary problems also arose because the corals were important in reducing wave activity at the coast, and their removal led to increased wave activity and coastal erosion.

Although storm hydrographs are very much a fluvial concern, the example from Kanehoe Bay shows that we are in a situation to observe potential problems for coastal areas. A sharp peak in the hydrograph (Figure 6.1b) means that there is a large volume of water moving down the river over a short period of time, effectively like a wave form. As a result, the water will have high energy levels and move large quantities of sediment, which will be deposited only when the energy levels of the river have dropped below the movement threshold. One place where this velocity drop happens is in the estuary or nearshore region where the river water meets the comparatively static water body of the sea. This process was well exemplified by the East and West Lyn rivers in north Devon, south-west England, which, over the period of a few hours on the night of Friday 15 August and the early morning of Saturday 16 August 1952, completely refashioned its estuary and part of the coastal embayment with sediment brought down by the floods which hit the two rivers owing to high rainfall over the previous day. In addition to the large volumes of sediment, the floods also did great damage to the town of Lynmouth, situated on the estuary, washing away many buildings and causing the deaths of thirty-four people, some of whom were never found (Delderfield, 1981).

While this sort of devastation may be considered dramatic, and a rare event, it does reflect the chain of events which can happen following heavy rainfall on a catchment with large areas of bedrock. The events at Lynmouth were natural, in that the catchment was undeveloped. However, increasing urbanisation and the concreting over of land can, in effect, artificially produce a catchment whose hydrological properties are similar to those of the East and West Lyn rivers.

Perhaps more common as a result of catchment modifications and changes to hydrographs is an increase in the frequency of coastal flooding. Any storm will increase the discharge of rivers, but the more peaked the hydrograph, the greater that discharge will be and the shorter the time for which it will last, and it is in estuaries that effects may be greatly felt. If such increased discharge reaches an estuary at low tide, then there is plenty of surplus volume in the estuary to accommodate that water, and there are few problems. If, on the other hand, the increased discharge reaches the estuary at high tide, when the estuary is already full, the only way in which the increased volume can be accommodated is for the estuary to overtop its embankments. Because most estuaries are the locations for settlements, this often causes coastal flooding and disruptions to settlement and infrastructure.

From a management point of view, such events are difficult to cater for. Better planning of catchment developments and reduction in concrete coverage may help, but at the coast itself, dedicated floodlands are a good way of ensuring the protection of settlement. Floodlands, as their name suggests, are areas of land which are allowed to flood in order to take excess storm water during periods of high water. They subsequently drain on the ebb tide. Thus they may be viewed as a form of artificial floodplain. Perhaps ironically, though, it is development on natural floodplains which often exacerbates the original problems.

In summary, it can be seen that coastal flooding can be increased as a result of development and land modifications in the catchment. In particular, the reduction in the amount of ground available for water to soak into means that more water flows over the surface, producing sudden and large flows of short duration down rivers.

Impacts of farming and land clearance

In addition to the catchment modifications mentioned above, other changes in land use may also have implications for the coast. Perhaps the greatest impacts are felt when forests are cleared for agricultural land. Natural soil loss in areas still forested has been estimated at up to 3 tonnes ha^{-1} yr^{-1} for tropical woodland. Forest clearance and cultivation increases this loss dramatically to between 54 and 334 tonnes ha^{-1} yr^{-1}. Even if the whole soil horizon is not removed, individual components may be, such as the fine material, leaving a very sandy and non-stable soil horizon, which is also very low in fertility. Simple mathematics will indicate the increase in available sediment, the bulk of which will enter the river system, with significant quantities being brought downstream to the coast. In addition to leaving the soil surface open to erosion, and hence increasing sediment loads, forest canopies intercept a lot of rainfall, and because trees store a great deal of water, their loss will increase the amount of water which runs off to the river system. It is not possible to do this without causing dramatic modification to the catchment. This occurs in several ways. First, increased run-off causes increased soil erosion owing to the lack of roots to bind soil and the increases in saturated ground. Second, greater volumes of water can cause the rejuvenation of rivers, which leads to increased erosion and so increased sediment yields. Third, the changes in storm hydrographs produce increased flow velocities for shorter duration, which again facilitates higher sediment loads.

All these processes point to larger volumes of sediment being brought down by rivers. This sediment can end up in one of two places. As the river enters its lowland channel, large amounts of sediment can be deposited as currents slacken. One place for this is in estuaries, where, owing to flocculation (see Chapter 2), large quantities of mud will settle. The silting up of estuaries in this way can have serious implications because it reduces estuarine volumes, which can lead to problems of navigation and threaten the existence of port facilities, and can also cause tidal flooding of settlements. In addition to the reduction in volume caused by sediment, the estuary also receives increased quantities of water from the river owing to increased run-off following land clearance, thus making the problem of flooding worse.

If the currents in the river are sufficient, then the sediment can be transported beyond the estuary out to the coast, producing a sediment plume or, exceptionally, a delta at the estuary mouth. This will represent a large sediment input into the coastal sediment budget, and can serve to benefit sediment accretion. Although this may be considered a positive benefit, detrimental effects can also occur, especially where there are fisheries or coral reefs which can become swamped, navigation channels which can silt up, or sandy beaches which can receive large quantities of muddy sediment

The engineering of rivers and the construction of dams

While the above example deals with increases in sediment due to land clearance, the reduction of sediment reaching the coast can also have severe implications. Such a reduction can occur in one of two ways. First, river management and engineering works to protect areas of erosion will remove that sediment source from the river sediment budget. This will mean less sediment moving downstream, although increased erosion elsewhere in the channel may counteract this. Similarly, sediment can be prevented from reaching the coast by the construction of dams or weirs across the river. As with barrages (Chapter 3), the building of an impenetrable structure across a river will impede discharge and sediment movement. In many cases of dam construction, there have been problems with reductions in sediment loads reaching estuaries.

The closure of the Aswan High Dam on the river Nile in 1964 led to significant reductions in the amount of sediment reaching the Nile delta. It has been estimated that pre-closure in 1964, the Nile input to the delta was of the order of 124 million tonnes of sediment per year (Stanley and Warne, 1993). Post-closure, this input had been reduced to 50 million tonnes per year (Carter, 1988). As we have seen from previous examples (Chapters 2 and 3), when the sediment input to the coast is reduced or ceases, then that part of the coast often starts to erode. In the case of the Nile delta, the removal of the river sediment source is the most probable cause of the large-scale removal of sediment from the delta by the sea (Orlova and Zenkovich, 1974) at rates of up to 30 m yr^{-1} at the Ras El-bar (Bird, 1985). It is important to remember in this example that the erosional regime of the sea has not changed. Before the dam closed, the sea still eroded the delta, but the eroded sediment was replaced by a greater quantity of river sediment, hence leading to a situation of net accretion. After closure, the sea has continued to erode in the same way, but the input of sediment to replace this erosion has stopped, and so the delta is now in a situation of net erosion. Other impacts have also been experienced in the Nile delta: in the fishing industry, and also in crop cultivation, owing to decreased fresh water and nutrient-sediment inputs.

Similar problems have been experienced in the Volta delta on the coast of Ghana (Ly, 1980). The Akosombo dam was constructed in 1961 and, since closure, has blocked around 99.5 per cent of the drainage from the Volta basin. The only sediment which now reaches the delta from the river is that which is eroded from below the dam, and is negligible in comparison to the original volumes of sediment brought down by the river. That sediment was originally the main source of sediment for the adjacent coastline. Although much of the coastline was experiencing erosion before the dam was constructed, the only areas to experience an increase in erosion since the dam was closed are those areas which formerly received sediment from the river. Retreat rates since the dam became operational have been in the order of an increase of 2–3 m yr^{-1} on pre-dam rates, with localised rates reaching between 8 and 10 m yr^{-1}.

Other deltas have also experienced similar erosion problems following dam construction and closure. Viles and Spencer (1995) report that the Ebro delta, Spain, has lost 96 per cent of its fluvial input, while the Rhône delta has lost 90 per cent. Clearly, such dramatic decreases in sediment supply to coastal deltas have severe implications for local and regional coastal stability. Lack of sediment input means that once-stable deltas become unstable and erosional. This will impact directly on land use, especially farming, which often predominates in such areas, but also, as in the case of the Mississippi, on settlement and capital-intensive land use.

Although dams represent major constructions in rivers, regulation of water flows by weirs can also have an impact, as even such small-scale features can retard sediment movement downstream and lead to sediment starvation at the coast. Reed (1992) demonstrates this with respect to a study of the impact of weir construction in a Louisiana salt marsh, which resulted in the marsh areas downstream of the weirs experiencing significantly lower accretion rates.

◆   ◆   ◆

We have seen three ways in which catchment-based operations can have impacts on the coast. It has not been necessary to describe the processes in detail, as this verges on the realms of river management, which are beyond the scope of this text. It does serve to illustrate the point, however, that the coast cannot be considered in isolation. Where coastlines receive their greater sediment input from rivers, then such rivers and their catchments become of great importance in the coastal context, because actions some distance up-river can lead to significant modifications at the coast.

These modifications can be summarised in three ways: by decreasing the time taken for storm run-off to pass through the system; by increasing the sediment yield by changes in land use; and by decreasing the sediment yield by blocking river flow. To some extent, these latter two are mutually exclusive and tend to suggest that there is a no-win situation. This is not the case, however, because the effects are also partly dependent on the river itself. If a river has a low sediment yield, and inputs little to the coast, then constructing a dam will not have any large-scale impact of sediment at the coast, as the coast is not reliant on the river as a prime source. Similarly, increased sediment yield on a river where there is already an abundance of sediment may not cause the same intensity of coastal problem because the estuary is likely to be sediment choked already.

Offshore causes

We have already seen how human impacts within the river catchment can cause problems at the coast. In a similar way, actions offshore can also have detrimental impacts. In this case, however, the problems are related less to sediment than to pollution. With regard to sediments, the offshore zone represents a long-term store, and little human intervention goes on relative to this. There are occasionally schemes, however, to utilise offshore sand and shingle banks for sediment sources, which could potentially lead to detrimental impacts at the coast.

Utilisation of offshore sediments

We have already explained that as far as coastal processes are concerned, deposits of sediment are merely temporary features, to be used once the accretion/erosion regime of the coastal system changes, and should not be seen as permanent. Similarly, while deposited, such features provide important offshore wave breaks, thus providing a natural analogue to the construction of offshore breakwaters (see Chapter 3). Bearing both these facts in mind, we see that any utilisation of offshore sediments needs to take into account the impacts for the coast. Removal of sand and shingle can leave the coast exposed to increased wave attack, while any development of these features renders them immobile as future sources of sediment.

We have already seen in Chapter 1 how the removal of offshore sediment structures can lead to increased wave erosion at the shoreline (Hallsands, Devon; Job, 1993). In a similar way, the many sandbanks off the coast of north Norfolk in eastern England serve to provide sufficient coastal protection from waves to allow the development of spits and salt marshes. In addition to providing protection, these banks also focus wave energy onto particular parts of the coast, thus causing a series of erosion and accretion cells. This further highlights the importance of offshore deposits in controlling what actually happens at the coast. Such features are mobile and do change their shape and size according to marine conditions. Any plans to develop such features, therefore, may seem poorly based in science. Plans do exist, however, to develop offshore sandbanks for features such as wind farms.

Closer inshore, Carter *et al.* (1992) demonstrate the environmental impacts of sediment removal from beaches. Using an example from Northern Ireland, they claim that 80 per cent of the area's coastal erosion can be accounted for by sand loss via extraction. This loss is due partly to beach lowering, and partly to the inability of dunes to store enough sediment to compensate for winter lowering of the beach profile. Such problems are not restricted to Northern Ireland, however. Carter *et al.* also indicate problems of varying magnitude in Australia, South Africa, Malaysia, Morocco, Tanzania, Namibia, Canada, the USA, Spain, India, Sri Lanka, and Brazil. In Scotland, sand mining occurs on 20 per cent of beaches because of the high demand for building aggregate, while on the Isle of Man (Joliffe, 1981), historic mining is claimed as the reason behind the island's current beach erosion problems.

The problems of shipping and accidents

The greatest offshore threat, however, comes from pollution. Despite many increases in ship safety, and tight regulations concerning routes and management practices, shipping accidents still occur. While oil spills receive the greatest reporting, other cargoes can cause damage to coastal environments. Despite this, perhaps oil is the greatest threat both because of its range of impacts and because of the length of time for which it can remain environmentally damaging. When oil pollutes a coast, it can kill the majority of life forms, both birds and infauna. The impacts on coasts can be dramatic and, depending on wave and tidal conditions, can cover large areas. Table 6.1 details the major oil spills since the mid-1960s and the extent of coastline contaminated, where known.

TABLE 6.1   Oil tanker accidents since 1967, and their impacts on the coast

Date	Ship	Location	Spill (tonnes)	Coastal impact[a]
1967	Torrey Canyon	Scilly Isles (S.W. England)	118,000	32 km contaminated, up to 100,000 birds killed
1970	Arrow	Chedabucto Bay (Nova Scotia)	10,000	300 km contaminated
1972	Sea Star	Gulf	115,000	—
1975	Jakob Maersk	Oporto	80,000	—
1976	Haven	Italy	40,000	—
1976	Argo Merchant	Eastern USA	28,000	—
1976	Showa Maru	Singapore	70,000	—
1976	Urquiola	La Coruña (Spain)	108,000	—
1978	Andros Patria	Spain	20,000	—
1978	Amoco Cadiz	Portsall (France)	223,000	300 km contaminated
1979	Burma Agate	Texas	42,000	—
1979	Atlantic Express	Venezuela	300,000	—
1983	Castillo de Belver	Cape coast (South Africa)	200,000	32 × 5 km slick
1989	Exxon Valdez	Blight Islands (Alaska)	10 million gallons	1,770 km contaminated 5,500 otters, 200 harbour seals, 400,000 birds killed
1989	Kharg-5	Gibraltar	82,000	—
1991	Haven	Genoa	140,000	—
1992	Aegean Sea	La Coruña (Spain)	72,000	—
1993	Braer	Shetland Islands (Scotland)	85,000	Many important bird communities damaged
1996	Sea Empress	Milford Haven (Wales)	70,000	190 km contaminated

Source: Compiled from Barrow, 1995; Daily Telegraph, 1996; Pearce, 1996; Pickering and Owen, 1994
Note:
[a] A dash in this column denotes that no information was available, not that no coastal impact occurred – which, in the majority of cases, it certainly did.

While impacts are immediate, clean-up operations can reduce the impact considerably. The use of chemical sprays does disperse the oil but just makes it sink to contaminate the benthos and bottom sediments. As oil is organic, it will break down naturally given time, although the problem remains that when it is beaches and valuable conservation coasts which are contaminated, it is important to remove the oil as quickly as possible. In cold climates, oil can remain in the system for many years. Hence, when incorporated into beach sediments, the oil is going to be broken down only very slowly. The *Exxon Valdez* spill in Alaska occurred in a cold climate, and so the natural regeneration of the area, and the breakdown of oil, will not occur as rapidly as would happen in warmer areas, hence the effects of the spill will be around for much longer.

Offshore industry

Oil accidents and spillages are rare events. Another aspect of the oil and gas industry which can also have an impact on coastlines is the actual extraction side of the operation. Such complex offshore activities need land-based support, and so designated oil ports undergo rapid expansion to service and support the rigs and other offshore operations. Similarly, where oil and gas are pumped directly onshore, storage areas are needed. Again, these structures require large areas of land which can become contaminated over time, and because of the contamination issue and the need to site installations according to the location of offshore operations rather than according to coastal suitability, these do not always occur on the most suitable coastlines. The Easington gas terminal, for example, is sited at the southern end of Holderness, on a coastline which is rapidly eroding.

In other respects, the very extraction process can also be unsound from a management perspective. Oil and gas extraction in the Mississippi delta, for example, involves the setting up of infrastructure and rigs on land which is best described as swamp. Because of this development, the whole coastal management and defence strategy for the coastline is brought into question. From a process point of view, the area is clearly not suitable for development, but because of the financial aspect of the area's reserves, these concerns have overridden those of the coastal managers and environmentalists.

In addition to unsuitable development, oil leaks from production platforms can also occur. While Table 6.1 details spills from shipping accidents, the world's worst oil spill occurred in the Gulf of Mexico (Hinrichsen, 1990), when in June 1979 a blow-out occurred in the Ixtoc 1 exploratory well in the Bay of Campeche. Between then and March 1980, when the well was finally capped, 475,000 tonnes of oil were spilled into the Gulf waters. In this case, the extent of the damage to the coast was never calculated, but the damage was considerable over a great distance (see Hinrichsen, 1990).

Another contentious issue, and a further possible threat to the coastal environment, concerns the disposal of the drilling platforms and storage containers. Exploration and drilling platforms do not pose too great a threat in this respect, as they are largely devoid of any stored oil products, and are generally designed to float into and out of ports. Offshore storage platforms, however, are more complex, and pose potential threats. Box 6.1 discusses this problem in more detail with reference to the Brent Spar.

Box 6.1 Disposal of the Brent Spar: arguments for and against land-based salvage

The debate on where to dispose of the Brent Spar hit the media headlines in the summer of 1995, when Shell announced that it wanted to dispose of its oil storage platform. The original intention was to dispose of the structure in around 180 m of water in the North Atlantic, some 240 km west of the Hebrides. Greenpeace objected to these plans on the basis (later shown to be erroneous) that large volumes of oil and sludge would pollute the sea bed, and that, in principle, the sea floor should not become a dumping ground for North Sea rigs, especially considering that there are several hundred other rigs awaiting decisions on disposal. Following additional pressure from Denmark and Germany, with threatened boycotts of Shell products, Shell abandoned its original intentions, indicating instead that the platform will be brought ashore for salvage. This pleased Greenpeace and environmentalists, who saw land-based disposal as a victory for the sake of the oceans, without thinking through the consequences for the land.

Clearly, disposal in the oceans needs to be regulated. However, it is important to weigh up the pros and cons of this method.

Arguments against ocean disposal:

- The old oil and ballast stored within the tanks could leak and contaminate part of the deep ocean.
- The structure will decay and possibly break up on the ocean floor.
- It would set a precedent for future disposal activities, and may open the door to other North Sea structures being dumped in this way.

Arguments against onshore salvage:

- The Brent Spar is over 137 m in height, with a draught of over 100 m. There is no port facility into which this could be towed for break-up, unless the structure was tipped on its side. This is the method originally used to launch it. Tipping it on its side can cause spillage.
- The ballast and oil would need to be pumped out and contained, prior to land-based incineration.
- The tanks, etc, would need breaking up, involving men entering them in hostile environments.
- All metal, etc., would either need to go to landfill, or be processed.

On the whole, there are great risks in bringing the structure into port. The process of tipping it could spill the contents. Manoeuvres would be difficult, and unfavourable conditions could cause further wave damage. In addition, two of the storage tanks are already ruptured, with the result that sea water has entered. These weaknesses make the structure even less stable, and possibly in danger of collapse when tilted.

On the whole, if land-based salvage were to be a success, Greenpeace and other environmentalists would score a moral victory. In contrast, the threat of coastal contamination is a risk which few port authorities would be willing to take. This brings in yet another problem: which port authority would allow such an operation to occur in its jurisdiction? What political pressures can local populations exert on councillors as part of the NIMBY trends?

The arguments against trend-setting are fully justified. It is also true that there are many structures in the North Sea which will need disposal. However, the Brent Spar is a one-off, there are no other similar structures present, and so the dumping of this type of structure will also only be a one-off event.

In effect, the Brent Spar represents a form of coastal protection issue that should form the basis of a one-off case. The potential threat to the coastline through contamination is great. It is perhaps safer to opt, in this case, for the easy way out, which is also, if all things are considered, the best environmental option; that is, dispersal at sea.

◆　　◆　　◆

It should perhaps also be said, for the sake of completeness, that leaks and spillages from production or storage installations which have direct impacts on coastlines need not be accidental. In the Iran–Iraq war, damage to the Nowruz Oil Field resulted in the release of around 6 million barrels of oil into the Gulf environment. In 1991, the Gulf war resulted in the release of a further 7–11 million barrels because of damage caused by retreating Iraqi forces. Coupled with this, a further 6–8 million barrels were released between 19 and 28 January 1991 from sunken and leaking vessels, including Iraqi oil tankers; and also from leakage from the Kuwaiti Mini Al-Ahmadi Sea Island terminal, and the Iraqi Mina Al-Bakr loading terminal (Alam, 1993).

During the Gulf war, the large oil releases formed a substantial oil slick. The movement of this slick was southwards, down the Saudi Arabian coast where winds continuously drove it onto the shore. The most heavily affected areas were the salt marshes and bays located between Ras Al Khafji and Abu Ali in the south. The environmental impact of this slick was great, causing the deaths of around 30,000 marine birds, the oiling of 20 per cent of the mangroves on the eastern coast of Saudi Arabia along with hundreds of square kilometres of seagrass and mudflats, and damage to around 50 per cent of the coral reefs. Other forms of damage also occurred in the region during this conflict, notably the trenching of beaches, vehicle movements, and the dozens of ships which were sunk in the Gulf. In addition, the destruction of the sewage treatment plants in Kuwait led to the release of 50,000 m^3

day^{-1} into Kuwait Bay, threatening the ecosystem, reducing water quality and polluting public beaches (Gerges, 1993; Price et al., 1993).

Long-term damage to the coastal environment has not been as serious as originally feared, however. Evaporation studies on oil suggest that winter evaporation can be double that of the summer, meaning that there was rapid evaporation of the oil spills, as they tended to occur under winter conditions. The rapid decomposition of oil meant that it had relatively short residence times. Subsurface oil was more persistent, with oil-saturated water common in many of the intertidal habitats.

Oil contamination of subtidal sediments was greatest in sheltered muddy bays (500–900 per thousand in some places). Many areas had escaped the large-scale problems of sinking oil, partly because of the fact that widespread use of dispersants did not occur, but also because much was carried on the surface. It is thought that much of what was deposited at the bottom was deposited there because it had been reworked from intertidal sources, and transported there as a result of sedimentation.

The penetration of oil into sediments was also a slight problem. Under water, penetration was along burrows, but on beaches, oil flowed through the sand grains. Where sands were coarse, the greater interstitial spaces allowed greater penetration, and hence greater contamination to depth. Where these beaches are sheltered, this oil has the potential of being around for some time. In areas of erosion, however, up to 70 per cent of the oil was remobilised into the subtidal zone, where it was deposited as tar balls. Overall, however, contamination appears to be restricted to the top few centimetres, and as a result, pollution effects are less severe than if the sediment was bioturbated.

A survey carried out a year after the spill revealed that while there was evidence that much of the oil had degraded, there were also areas which appeared fresh and were being supplemented by continuous releases from the sediment. A further survey by the International Council for Bird Preservation revealed that while no bird species had its world population reduced to such an extent that recovery was impossible, locally 20–50 per cent of species experienced severe mortality. In addition, the intertidal flats of the Saudi Arabian coast, important feeding areas for migratory birds, were heavily polluted, with a drastic reduction in the numbers of birds supported. In total, around 50 per cent of the mainland Gulf coast of Saudi Arabia was polluted with oil, as well as the shores of two offshore islands. Hence, the impact on wildfowl was significant.

A second survey, carried out in 1992, revealed that damage done to many of the coral reefs had been largely superficial, despite the oiling which they had received. Locally, however, three reefs in Kuwait showed serious signs of stress, typified by bleaching and mortality. It is possible that these effects on the corals are not due to the oil spills, but reflect the fact that the winter of 1991/92 was around 4–5°C colder than normal. It is also possible that the corals were subject to some form of pollution effects, notably from oil spill, smoke clouds, and sewage/pollution releases. On the Saudi reefs, there were no detected impacts from pollution. It is not possible to say definitely, however, that there will be no impacts. The reef communities have long histories of fluctuating populations, and so it is possible that the effects of the war are being masked. Long-term monitoring will help to reveal any overall trends.

Colonies of seagrass appear to be thriving, despite the fact that they had undergone complete covering in oil. Algal mats suffered extensive damage in the upper intertidal zone. The ability of a species to survive appears to be linked to the geomorphology and persistence of oil. The greatest impacts were on communities at the head of bays, where there was typically no sign of life on the mid- to upper fore-shore. The deep penetration of oil into burrows is likely to result in the continued pollution of these areas for many years to come.

There was evidence of problems and mortality in fish stocks, but there are other factors which could also have caused this effect. In some areas, the links between the oil and damage were slightly more conclusive. The World Conservation monitoring centre suggests that there is damage in heavily oiled areas, which includes long periods during which fishing was not possible. Surveys of fish populations have indicated that problems have been experienced locally. In some areas, there have been no reported losses, while in others, no fish can be found. The Saudi fisheries company reports that shrimp production is now less than 1 per cent of pre-war levels, although there may be more reasons to this than just pollution. Reproduction rates are also very much lower; this fact may also owe something to the problem of water temperature, because the breeding season is generally in the spring period, when rises in sea temperature occur. In 1991, this period occurred at the same time as the appearance of the smoke plume, which decreased atmospheric temperatures significantly (sources: Alam, 1993; Gerges, 1993; Gupta *et al.*, 1993; Price, 1993; Price *et al.*, 1993; Watt *et al.*, 1993).

Summary of Chapter 6

This chapter has been a brief introduction to some of the wider issues relating to human impacts at the coastline. It is important to realise that a particular environ-ment can suffer from problems which have their origins some distance away, these problems being largely sedimentological and pollutant based. We have already seen that many of the problems at the coast have their origins in lack of sediment. Sediment variation from a river acts in exactly the same way as any sediment source at the coast.

Pollution-wise, many coastal habitats and the wildlife which inhabits them are vulnerable to pollution. We have concentrated here on oil, which is perhaps one of the most damaging pollutants. The sources are also varied. Accidents do happen. All management can do is to try to put rules and regulations in place to minimise them. This may involve locating dedicated shipping routes away from sensitive coastal areas, but this cannot always be possible. Milford Haven in Wales, for example, is a prime oil terminal, but is surrounded by seventeen designated sites of special scientific interest and two marine nature reserves, and experienced severe oil pollution in 1996. In this case, however, it has subsequently emerged that tidal current surveys were incomplete.

One of the chief coastal oil pollution incidents of this decade has been the Gulf war. What happened there was indefensible, and marked a new era in eco-warfare. Militarily, nothing was really gained from the exercise of setting fire to oil installations and blowing up wells. Environmentally, however, the cost was high. Because of the nature of the intertidal environment here, the spilt oil will be reappearing in the environment for many years to come.

On leaving this chapter, readers should realise that not all coastal issues have origins which are related to the coast and its processes. Any activity on the land can influence sediment and water flow through rivers. These can all impact on the coast. Many sea-based operations can equally impact on coasts through pollution. Pollution incidents are hard to manage from a coastal management point of view. The best that can be achieved is the provision of emergency plans to cope with problems once they arise. Land-based problems, however, can be accounted for in catchment management plans. What we have seen in this chapter is that from a coastal point of view, these should not be ignored because there are coastlines which owe their stability to rivers.

Management frameworks for coastal and estuarine systems

Introduction

The aim of this book has been to investigate the ways in which human intervention at the coast has interfered with natural processes. While many of the original problems may have been solved, this interference has caused further problems along the adjacent shoreline. We have constantly seen how engineering methods, whether hard or soft, have been used to protect a part of the coast from erosion, either by building up the coast with sediment, or by constructing a physical barrier between that which causes erosion, and that which is being eroded. Clearly, then, humans have the capability to fix the coastline into its present form; all it needs is enough money to pay for the work and to maintain it.

While we clearly have the relevant knowledge to protect our coasts, and to safeguard the hinterland, it is all too common to do this from a very narrow viewpoint. It is only fairly recently that coastal engineers have taken a broader view and looked further afield, and not just concentrated on the piece of coast which is to be defended. By looking both up- and down-drift you start to see what is happening in the wider context, and, most importantly, what is happening to the sediment. Doing this makes it easier to appreciate the impacts of engineering structures on the rest of the coast. Table 7.1 details the increasing perspective which such a process can provide. It becomes quite clear that the broader the view taken, the more of the coastline you look at, meaning that you can gain a greater amount of information regarding how that coastline operates. Just addressing the problem of the immediate area to be defended takes no account of the impacts in adjacent areas, and so makes it impossible to predict what may happen to beaches down-drift, to sediment transport patterns, or to sediment source areas and the sediment budget. Clearly, this sort of approach is very restrictive, and is liable to cause more problems than it solves.

Each successive scale change (Table 7.1) increases the amount of available information, and so can lead to safer predictions with regard to the impacts of any defence works. At this broader scale, however, we start to obtain a lot of information from many different sources. In addition, we also begin to draw in many different coastal types, land uses, and activities. This process ultimately leads into the concept of managing the whole coast. By taking a more holistic view and observing how the coast operates, it becomes possible to identify areas where certain activities may be more suited, where the construction of defences would have serious impacts on coastal sediment budgets and where it would not, and where development of the hinterland would be unwise owing to current or predicted erosion trends.

It is at this point that we return to the recurrent theme which has become more and more evident as this book has progressed. It is only by viewing the coast as a single unit that it can become possible to use it in a sustainable way. In order for this

TABLE 7.1   The increased environmental information available to the coastal planner as a result of increasing the area of coastline used for data-gathering

Area of section	Information available	Prospective management unit
Area of defence	• Sediment sizes • Coastal morphology • Types of structure needed • Impacts on local beaches • Hinterland/recreational uses	Sea/flood defence structure
Area of defence plus immediate upstream/ downstream area	All the above plus: • Impacts on local wave climate • Sediment source areas • Impacts on neighbouring beaches • Areas affected by edge effects • Area of sediment starvation	Sea/flood defence structure
Urban frontage	All the above plus: • Overall defence strategy for coastal section • Pattern of land use and values • Some ideas about areal coastal management	Coastal management plan
Embayment	All the above plus: • Management of coarse sediment budget • Sediment sources and sinks • Sediment pathways • Wider impacts of defences	Coastal sub-cell
Coastal unit	All the above plus: • Macro-scale sediment transport and deposition patterns • Wind/wave climate • Offshore sediment sinks	Coastal cell
National coastline	All the above plus: • Offshore impacts	The coastline of a country

to be the case, we need a management scheme for each part of the coast. The realisation of this need, however, has come about in completely the wrong way. It is because coastal engineering structures have caused major problems along the world's coastlines, deriving from their restrictive and rather piecemeal approach, that many people have begun to realise that some form of overall management would offer new solutions to both natural and anthropogenic problems. Engineers have now realised that the coast is a complex system operating over a large area. What is done in one part of this area can produce effects in other parts. It is only from the realisation that coasts operate in this complex way that the delineation of coastal cells has been possible, thus putting in place the basis of the coastal management framework.

This increase in realisation has been paralleled by an increase in defence construction which has resulted in the isolation from coastal processes of more and more sediment sources. This, in turn, has produced the need for further defences. Clayton (1990) has demonstrated this phenomenon with the growth of defence structures in the UK over the past century. Furthermore, we are also beginning to recognise a new problem. Many of the existing defences are quite old, especially in established tourist areas where defence structures were constructed a hundred or so years ago, and are nearing the end of their useful life. This fact poses a new threat to the security of coastal areas, because even though an area is defended, poor defences are only marginally better than no defences at all, and so many areas may be resting under a sense of false security. The fact that many defences are due for replacement does, however, provide coastal planners and managers with a good opportunity to rethink the coastal strategies of a stretch of coast. As defences become due for replacement, it is possible to reassess their viability and usefulness by asking whether the area in question still needs defending. Can it be better defended by soft engineering? If we free up sediment movement along the coast, can it protect itself by habitat generation?

If we also consider the problems of sea level rise (discussed fully later in this chapter), then it becomes advantageous to free up as much of the coastline as possible. The option of a more natural coastline can only lead to future benefits, because the more sediment that can be freed up now, the more natural habitats can develop in the future, and so the coast can respond to changes in wind, waves, and sea level more easily. Naturally, nobody is saying that such options are universally appropriate. A major tourist resort depends for its survival on defences, and so in such cases it is important to maintain the line of defence. Similarly, the Netherlands would take on a whole new outline were it not for its defences. Clearly, then, coastal management has to take account of all these views. As far as coastal defence is concerned, should new defence replacements hold the existing line, advance seaward, or retreat landward?

Coastal management is not only about defence, however. Coastal use is also an important part but, while some uses, such as tourism and ports, are legitimate uses of the coastal margins, many other developments are not, and need not necessarily be located in these areas. Similarly, with resort areas, should the development of these sites be allowed to continue unchecked, as has happened in the Mediterranean (Chapter 5), or should limits be placed on their size and the type of development allowed?

We appear to be entering a wide-ranging debate here, and one which has about as many answers as it has questions. There is a general agreement that because of the increased knowledge of how the coast works and how different uses can impact upon it, it is necessary to adopt a holistic form of coastal management. Only by looking at the coast as a whole can we produce a sensible management strategy. This, therefore, represents a major new sphere in the realm of human interference at the coast and in estuaries. It is because of this growing field that this final chapter deals in detail with the subject of coastal management. We have seen how human activity has had many negative impacts; we will now see how humans are beginning to adapt their approach to coasts to be more user-friendly, and to try to address some of the problems caused.

The concept of coastal zone management

The tendency for coasts to be used by many different sectors of society has led to a proliferation of controlling and interested groups when it comes to coastal management. Because of the problem of sheer size, the management of many coastal areas is divided into smaller, more manageable units, each with its own management plan. These plans tend to form a hierarchy, in which plans for smaller units join to form plans for larger coastal units, and so on. Table 7.2 details the types of management strategies which occur along the coast.

While the development of estuarine and coastal protection plans is important on the local scale, there is still a potential problem of activity in one area having a detrimental effect on processes in another. It is essential that these local plans are linked to a broader scheme based on regional and national strategies and management. The common denominator in all this is the formulation of a national coastal defence strategy based on coastal zone management plans (Table 7.2). In its purest form, the concept of coastal zone management (CZM), also referred to as integrated CZM (ICZM), involves an ecosystem approach to the problems of management. Any human activity or development along the coast must be undertaken only with due consideration of the environmental needs of erosion, accretion, and habitat creation and stability. Increasingly, however, the term has been applied to the whole field of

TABLE 7.2 The relationships of statutory and non-statutory management initiatives at national, regional, and local levels

Level	Non-statutory management	Statutory management
National (coast)	Policy initiatives Shoreline management plans Regional strategies	National planning policy Structure plans
Regional (coastal cells)	Coastline protection plans Coastal cell strategies Coastal zone management plans Coastal management plans	Regional development plans Regional policy guidance Local structure plans
Local (coastal sub-cells)	Estuary management plans Harbour management plans Catchment management plans	Tourism management plans Development plans

coastal activity such as the management of competing land use activities: fishing, versus shooting, versus sailing, for example.

So, using the modern definition of the term, coastal zone management can be seen as a process which covers the whole concept of protecting the coastline from damage and change from any activity. The topic itself has been the feature of many texts (such as Beatley *et al.*, 1994; Clayton, 1993; Goldberg, 1994; Healy and Doody, 1995; and Organisation for Economic Co-operation and Development, 1993), and a detailed discussion is beyond the scope of this book, although it is important to consider some aspects of the topic as a tool with which to moderate the human impacts on the coast. There are many ways in which the CZM process can be summarised, although as a general rule we can say that CZM represents a dynamic process which develops and implements a co-ordinated strategy to allocate resources to achieve the conservation and sustainable multiple use of the coastal zone.

Aims and requirements of coastal zone management

It is important when considering CZM that certain basic aims and requirements are stipulated. It is generally agreed that with regard to CZM, the coastline includes the hinterland, the intertidal area, and the subtidal areas. However, there is no overall view as to how much of the hinterland and how much of the sea should be included. Owing to the variation in coastal topography and coastal process, the precise landward and seaward limits are difficult to determine for a single CZM policy, and will vary according to the area in which the CZM is being adopted. Important factors in the consideration of areal coverage are the primary coastal processes – that is, erosion/accretion – and the primary coastal usage – that is, tourism, fishing. A CZM plan on a coast of severe erosion will take in more of the landward margin in order to control what developments occur near to the sea margins, thereby limiting the necessity for future defence measures and their impacts. In contrast, on a stable coast dominated by the fishing industry, more of the sea will be taken in, to protect spawning grounds, shellfish beds, and water quality. As far as these measures are concerned, the CZM scheme retains a degree of flexibility in recognition of the fact that all coastlines are different and the usage which each of them experiences will largely determine the form of management which they require. Juhasz (1991) supports this idea of variability in areal coverage of a CZM plan, and develops the idea further in the form of legislation, proposing the need for both broad-scale legislation to deal with problems such as erosion and accretion, and also more specific legislation dealing with aspects like pollution, shipping, and conservation.

The one important criterion, however, is that the basis for each scheme should centre around the environmental principles of sediment movement and habitats. By adopting this policy, and fitting socio-economic planning around it, the impact of any development on the environment, and vice versa, can be minimised. Thus development may be allowed in an area of coastal stability, provided that stability is predicted to remain, but not in areas of instability, thereby reducing the need for further defence measures. However, before a CZM plan can be instigated, it is important to know the detail for this environmentally based designation. This detail will involve defining the area that the plan will cover.

Defining the landward boundary

Bear in mind that while environmental criteria are paramount, the CZM scheme also has to function in a workable manner, and so it is necessary to impose some constraints on the plan which may not be purely environmental in their consideration. The main problem with defining how far inland the CZM should go is the limit that one should choose: whether one based purely on environmental grounds or one which coincides with an existing administrative boundary. It is very rare that the two coincide.

As has been said before, the best approach will be one based on environmental grounds, such as:

1   influence of maritime climate, that is, the limit of salt spray, hence including coverage of all habitats which are under the maritime influence (intertidal and supratidal), including cliff-tops and sand dunes;
2   mean high-water mark, that is, all intertidal habitats, but excluding supratidal areas (dunes and landward transitions of salt marshes and mangroves);
3   boundary of watersheds; that is, the whole of the catchment of rivers entering a particular coastline. This policy will include coverage of all activities which can have an effect on the coast; that is, dams and their effects on sediment loads, land management such as deforestation and the effects on sediment mobility and input, or industry and the increase in pollutants;
4   an arbitrary limit set according to the conditions in any specified area taking into account local topography, land use, and habitats.

All the above are idealised situations. Clearly, if the management plan wants total control, then a scheme based along the lines of 3 above will cover all possibilities. In reality, however, this involves huge areas of land, and is, in practice, largely unmanageable, because in most countries all rivers enter the sea and so the coastal plans will cover the entire land mass. In reality, the distance adopted is an arbitrary limit, of the order of 2 km inland from the mean high-water mark of ordinary spring tides (MHWOST), modified according to the geography and issues relevant to a particular coastline. The problem of river catchments is subsequently included by reference to catchment management plans (Table 7.2) (see Chapter 6), which are developed to aid the management of river habitats.

Defining the seaward boundary

Because natural marine processes recognise no administrative boundaries, designating seaward limits becomes difficult. Ultimately, the limit of national jurisdiction may be the obvious choice, but in reality many processes which influence the coast (shipping, fishing, etc.) occur outside this limit. As with the definition of the landward limit, there are a series of other possibilities for the seaward limit, which include:

1.   mean low water mark; that is, intertidal areas but ignoring subtidal areas – which means excluding marine interests and coral reefs;

2.   edge of the continental shelf; that is, the limit of sediment stores, but this does cover large areas in many cases. Also, some countries, such as Germany, the Netherlands, and Denmark, do not have an edge to their shelf;

3.   bottom of the continental slope for similar reasons as for the edge of the shelf, but including the loss of sediment to the deep oceans via sediment avalanches (turbidite flows) down the continental slope onto the abyssal plains;

4.   an arbitrary water depth, although this can be quite variable in areas of high sediment mobility (formation and removal of sand flats), which is itself an important consideration in management. In addition, water depth can increase and then decrease again, making such a limit difficult to delineate;

5.   one of the major, internationally acknowledged administrative boundaries. In the UK, there are three such boundaries:

      (a)   3 nautical miles – limit of controlled waters;
      (b)   6 nautical miles – limit of exclusive fishing rights;
      (c)   12 nautical miles – limit of territorial waters;

In this case, the favoured limit is the 12 nautical mile mark, as this covers most of the sea activities over which a country has some jurisdiction.

Methodology behind coastal zone management

We have consistently seen how human impacts on the coastal zone have caused knock-on effects for natural processes and coastal stability. Coming out of this is the growing realisation that if left to its own devices, this intervention will only produce further conflict between different user groups and between the natural environment and coastal protection measures. Eventually this will impact on the whole viability and stability of the coastline.

Many countries have realised the problems associated with management of the coast, and have undertaken initiatives to correct them, largely by constructing formal management policies. This has put CZM into the international arena, resulting in an increased awareness of many trans-boundary problems, such as pollution, dam construction and sediment supply, and tourism. Barston (1994) highlights the increased potential which an international arena provides, but demonstrates how this can also produce a need for improved information systems and the development of international standards. It is here that a further, and significant, problem arises. Once it is put on an international scale, coastal managers and other experts tend to lose control of the whole CZM process, with politicians becoming the driving and negotiating force. Similarly, many international problems can also arise, such as where one country's coastal defence policy can impact on another. For example, the implications of the UK's defending the eroding cliffs of eastern England will be felt in the Netherlands and Germany; or consider the breaking up of the Soviet Union and the associated policy problems which this caused, not least the problems concerned with how to redistribute fish quotas (Barston, 1994).

Despite these problems, many nations have developed their own CZM initiatives, producing a proliferation of examples and case studies. Owing to the vast number of examples cited in the literature, it is impossible to produce fair coverage of all countries here. Table 7.3 contains a diverse bibliography of the CZM experiences of many countries. By following these leads, and using the reference lists provided in these sources, the reader can easily develop case studies for any of the countries presented.

The UK appears to be lagging considerably behind in the production of overall coastal strategies. In 1992, the UK Department of the Environment published a paper entitled *Coastal Zone Protection and Planning* (Department of the Environment, 1992), which represents the first real advance for the concept of managing the UK coast as a single unit, composed of many individual but interacting parts. Following this, the Earth Summit in Rio de Janeiro in 1992 led to the production of Agenda 21 as a blueprint for environmental action. Chapter 17 deals exclusively with oceans and coastlines, and probably represents the strongest international commitment to CZM that currently exists. Agenda 21 commits coastal nations to the implementation of integrated coastal zone management initiatives, and to the sustainable development of coastal areas and marine environment under their jurisdiction. Within this, a number of co-objectives commit signatories to more specific aspects of coastal management:

1   To provide for an integrated coastal policy and decision-making process in order to promote compatibility and balance of coastal uses. This also stipulates the inclusion and co-operation of government departments, ministries and agencies which have control over specific aspects of the coast.
2   To apply preventive and precautionary approaches in development including prior assessment and systematic observation of the impacts of major projects.
3   To promote the development and application of techniques which reflect changes in value resulting from uses in coastal areas. These changes include pollution; loss of value due to erosion; loss of natural resources and habitat destruction, on which assigning a cost is difficult; and the increase in value due to development of the hinterland.
4   To liaise with all interested groups, to provide access to relevant information, and to provide opportunities for consultation and participation in planning and decision-making processes associated with the development of management plans.

These policies have also been supported from other sources. The Intergovernmental Panel on Climate Change (IPCC) recognises the need for management to allow for the continued protection of the coast in the light of global environmental changes and sea level rise.

TABLE 7.3   Select bibliography of coastal zone management initiatives in different countries

Country or area	Author(s)	Region
Africa	Ngoile *et al.* (1995)	Eastern
Asia (South-East)	Chou (1994)	—
Australia	Craik (1996)	Great Barrier Reef
	Crawford (1992)	—
	Cullen (1980, 1982)	—
	Hanley and Couriel (1992)	Northern Territory
	Harvey (1988)	Murray estuary
	Haward (1993, 1995)	—
	Hildreth (1992)	Queensland
	O'Brien (1988)	West coast
	Pearce (1991)	West coast
	Yapp (1986)	—
Bahrain	Almadany *et al.* (1991)	—
Bangladesh	Bashirullah *et al.* (1989)	—
Barbados	Miles *et al.* (1995)	—
Belize	Katz (1989)	Caribbean
	Mumby *et al.* (1995)	—
Bulgaria	Archer (1995)	—
Canada	Day and Gamble (1990)	British Columbia
	Georgison and Day (1995)	Vancouver and British Columbia
	Harrison and Parkes (1983)	—
Chile	Castilla (1996)	—
China	Chen (1989)	—
	Degong (1989)	—
	Lu (1990)	—
	Xu and Zheng (1991)	Bohai Sea
Costa Rica	Sorensen (1990)	—
Ecuador	Broadus and Gaines (1987)	Galápagos Islands
	Merschrod (1989)	—
	Olsen (1987)	—
Egypt	Eid and Fawzi (1991)	Red Sea
	Frihy (1996)	Nile delta coast
Estonia	Ratas and Purmann (1995)	Western archipelago
	Riggs (1994)	—
Europe (general)	Ballinger *et al.* (1994)	General European coastline
France	Guilcher and Hallégouët (1991)	Brittany
Hong Kong	Holmes (1988)	Tolo Harbour
India	Mallik (1987)	Kerala
	Nayak *et al.* (1992)	—

TABLE 7.3   continued

Country or area	Author(s)	Region
Indonesia	Claridge (1994)	Jambi Province, Sumatra
	Sloan and Sugandhy (1994)	—
	Soegiarto (1981)	—
	Yates (1994)	Kepulauan–Seribu marine park
Israel	Amir (1984)	—
Italy	Bettinetti *et al.* (1996)	Venice
Japan	Fluharty (1984)	—
	Shapiro (1984)	—
Korea	Hong (1991)	—
	Hong and Lee (1995)	—
Kuwait	Abouseida and Alsarawi (1990)	—
	Albakri (1996)	—
	Alsarawi *et al.* (1996)	Failaka Island
Mediterranean (general)	van der Meulen and Salman (1996)	General Mediterranean coastline
Netherlands	van Dijk (1994)	North Sea coast
	Koekebakker and Peet (1987)	North Sea coast
New Zealand	Flood *et al.* (1993)	—
	Haward (1993, 1995)	—
	Hickman and Cocklin (1992)	—
Philippines	Iriberri (1984)	Puerto Galera
	Yap (1996)	—
Samoa	Templet (1986)	American Samoa
South Africa	Kalicharran and Diab (1993)	Isipingo Inlet
	Sowman (1993)	—
South America	Sorensen and Brandani (1987)	All Latin American coastlines
Spain	Montoya (1991)	Mediterranean
	Palanques *et al.* (1990)	Ebro delta
	Suárez de Vivero (1992)	—
Sweden	Grip (1992)	—
Tanzania	Horrill *et al.* (1996)	Mafia Island
Thailand	Tabucanon (1991)	—
Turkey	Ozhan (1996)	—
UK	Earll (1994)	—
	Fuller and Randall (1988)	Orford Ness
	Houston and Jones (1987)	Sefton
	Morgan *et al.* (1993)	Glamorgan, Wales
	Roberts and McGown (1987)	North Kent
USA	Archer and Knecht (1987)	—
	Barley (1993)	Florida Keys
	Day and Gamble (1990)	Washington, Oregon, California

TABLE 7.3   continued

Country or area	Author(s)	Region
USA cont.	Fischer *et al.* (1986)	Florida
	Grenell (1991)	San Francisco Bay
	Guy (1983)	Florida
	Imperial and Hennessey (1996)	—
	Klarin and Herschman (1990)	—
	Knecht and Archer (1993)	—
	Knecht *et al.* (1996)	—
	Psuty (1983)	Eastern seaboard
	Walsh (1982)	—
USSR (former)	Bondarenko (1990)	Coastline of former USSR

Note: A dash in the 'Region' column indicates that the work refers to the general coast of that country, rather than to any specific region.

It appears, therefore, that all the guidelines and incentives are in place for countries to develop detailed coastal management plans for their coastlines. All these reports agree that it is essential to get away from the traditional approaches to management, which tend to be sectorially orientated and fragmented, and go for the management of the coastal zone as a whole unit. This problem is clearly exemplified by a study of the Fylde coast (Box 3.2, p. 77), where several maritime authorities along this coast each have their own form of sea defences. Clearly, if this coast were to be managed under the guidelines laid down by Agenda 21, these problems would not occur because the coast would be managed as a whole unit. This might involve the removing of groynes, or beach feeding the shore at Blackpool with the sediment extracted from Lytham.

The theories for developing a CZM scheme may be clearly laid out, but there are, nevertheless, some important questions. Most important are questions such as who does what? Who formulates the plans? Who enforces the plans? Who pays for the plans? This brings us on to the next important consideration: actually turning the concept into a workable framework and policy, which represents perhaps the prime area in which CZM ideals have faltered in many cases.

Running the CZM process

There are a whole series of groups that have a vested interest in the coastal zone. In many countries, these interests have been co-ordinated under one administrative authority, such as an environmental protection agency. In the UK, however, there is no single authority, just a collection of groups. These controlling groups have been appointed over time to administer the various acts and legislation which been introduced on a piecemeal basis to manage various aspects of coastal problems.

As there is no single group with individual control of the coastline, the policies of the groups involved can become fragmented. A good example of this is coastal defence policy. Until recently, the Coast Protection Act 1949 was administered by the Department of the Environment (DoE), while sea defences were provided for by the Ministry of Agriculture, Fisheries and Food (MAFF) under the Land Drainage Act 1976 – two similar and compatible responsibilities but administered by different parts of the government. Another aspect of this conflict is that the jurisdiction of the individual Acts depends on whether the land is above or below high water. Around the coast, there are many transition points where the level of land falls below or rises above the high-water mark. At these points, the relevant legislation changes from one Act to the other, and therefore from one department to the other. This was clearly a ludicrous situation, and the giving of responsibility for the Coast Protection Act to MAFF has opened the way for better strategic planning. Despite this, however, there are still conflicts with regard to flood protection (protecting property from inundation by sea water) and coastal defence (preventing land erosion).

Practical experience in many countries has shown that the best way to operate such a coastal management system is to have one group to oversee the whole operation. The favoured approach is to set up CZM 'units' to act as a stimulus and development focus within which all relevant areas of research can be focused and compiled to produce a comprehensive understanding of a particular coastal cell. At the present time in the UK, the area covered by individual CZM plans is administered by a range of administrative groups, meaning that the potential for fragmentation, non-co-operation and poorly constructed plans is higher than it would be were a single unitary authority in charge. By using the CZM ideas, a successful coastal management authority, on behalf of the people concerned in formulating the plan, should:

1   provide advice on CZM matters by facilitating contact between different groups and information sources;
2   provide a framework for CZM plans, in order to remove problems of over-lapping research, and to ensure comprehensive coverage of the study area;
3   co-ordinate and liaise between other groups working in adjacent coastal areas to stimulate discussion between people doing similar jobs, and also to maintain a consistent approach;
4   co-ordinate internationally with other CZM units;
5   seek and allocate funding to allow new research in areas of poor knowledge or understanding;
6   regulate plans for development via planning policies to oversee any projects which are likely to impact on the coastal management area;
7   implement CZM recommendations by making sure that they are known about and followed.

The requirements of CZM

With such a wide brief and so many contrasting demands on coastal resources, the interest of the CZM groups needs to be diverse. When the coastal zone is being planned, there are several factors which need to be taken into consideration, and included in overall plans. These factors form the basis of report structure and represent the main aims of the management plan. These are outlined below.

Protection from storms

Many coastal areas are in danger of damage from natural hazards. Hazards can take many forms, ranging from earthquakes, storms, surge tides, to sea level rise. Storms and hurricanes can cause sudden and dramatic effects on the coastline, causing the loss of life and property, and also have the potential to alter the coastal morphology by breaching sea defences. Although warning systems are becoming better developed, these cannot protect property, but as a means of ensuring public safety they can enable the evacuation of danger areas. In order to protect coasts from storms, there are a series of measures which could be implemented by a management plan.

1   Where the threat of storms is great, the protection of property may be too expensive. In such cases, the protection of human life is of prime concern and so it may be sufficient to arrange evacuation procedures in response to storm warning systems.

2   Where direct protection is feasible, actions such as reinforcing or increasing the height of sea walls may be preferred.

3   Particularly in estuaries, the main threat from storms comes in the form of a surge tide (meteorological tide). Estuaries which are prone to such features, or which contain significant areas of high-value hinterland, need to be protected by tide surge barriers, such as those found on the estuaries of the rivers Thames and Colne.

4   Where hinterland values are not that high, sufficient protection to property could be supplied through the provision of floodlands to take excess water. Nominating areas which are allowed to flood may enable more valuable areas to be protected.

5   The greatest impacts from storm activity are due to waves. We have seen (Chapter 2) how beach profiles can reduce wave action at the coast by shallowing gradually towards the land. By artificially building up beaches by beach feeding, sufficient wave attenuation can reduce the impacts of storms at the coast, and thus remove the need for increased hard defences.

6   Naturally, the best form of protection is prevention, hence one of the prime management recommendations has to be to avoid building in high-risk areas. Such planning controls can remove the necessity for any of the other methods mentioned here.

7   On coasts where people want to live, but which are subject to severe storms, such as in parts of the USA and Australia, planning regulations can be used to ensure the construction of 'storm-proof' buildings which can withstand the impacts of storm activity. However, it should perhaps be made clear to people that if, after information is made available, they still insist on developing

unsuitable areas, then they alone are responsible for the outcome, and should expect no protection to be forthcoming.

8   The impact of storms is also likely to become greater as a result of sea level rise. As sea level rises, wave base could deepen, thus increasing the force of a coastal storm. Bondesan *et al.* (1995) discuss this problem in relation to the Italian coast, and highlight the problem of predicted relative sea level rise (stressing the components of falling land levels and rising water levels) of 0.5 m by 2100 with respect to increased flooding from surge and storm tides. This makes coastal planning even more important, and indicates the need to identify areas susceptible to storm surge inundation.

EXAMPLE: THE THAMES BARRIER

The potential problem of sea level rise and the threat of storm surge down the North Sea led to the realisation that London was under serious threat from sea flooding. Because of the high population density, and the huge underground development in London, the threat of flooding from the Thames necessitated the building of the Thames Barrier. Despite the building of 112 km of defences in the nineteenth century, serious flooding took place in 1881, 1928, and 1953. With sea levels rising in the area, rises ranging from about 1.4 mm yr^{-1} at Southend to 2.7 mm yr^{-1} at London Bridge (Pye and French, 1993b; the difference is due to the influences of tidal wave amplification in the upper estuary arising from land reclamation and dredging – see Chapter 3), the danger from possible floods, particularly on surge tides, became high, and a realistic possibility following the 1953 surge.

The barrier was completed in 1984 and is designed to withhold a flood of up to 1.2 m greater than anything previously recorded. It has already been used on over twenty occasions following predictions of storm surges in the North Sea, effectively saving London from flooding.

Shoreline erosion

As we have seen, many problems of coastal erosion are brought about by the impact of humans on the coast. The dilemma facing the management team is whether to defend an eroding coast or not. If the area is undeveloped, it is likely that the coast will be left to erode in order to maintain sediment supply to the intertidal zone. When there are homes and livelihoods at stake, the decision becomes more complicated, as it depends on the value put on the land as to whether it is defended or not.

One of the problems with managing the coastal fringe is that many of the activities present have a good argument for being there, and so sensible zoning and multiple usage are generally the answer. However, it is fundamental that natural processes are given paramount importance, as without a strategy for defending the coast, many areas would be unusable owing to problems with instability and erosion.

As the successful functioning of the coast as a natural system is largely based on the movement of sediment from one point to another, any development which upsets this balance will cause a knock-on effect for other parts of the coast, and the setting up of a whole new set of management problems.

CZM has to reflect this dynamic nature, while protecting important areas. In order to maintain a balance between sediment supply (erosion) and deposition (accretion), the policy of CZM is to defend where necessary, but allow other areas to erode to maintain sediment inputs, even if this means the loss of isolated houses, small villages, or farmland (e.g. on the Holderness and Suffolk coasts of the UK). Only by making such sacrifices can greater protection and stability be given to other, more highly valued coastal areas.

Protection of coastal waters

Many coastlines are subject to a variety of pollutants from industrial, urban, or agricultural sources (Chapter 4). Extreme nutrient enrichment can cause algal blooms, causing water clouding and reduction in oxygen levels for marine life.

In the USA, many of these problems are dealt with through schemes known as best management practices (BMPs). For example, BMPs for agriculture include contour ploughing, crop rotation, filter strips, and animal waste control, to reduce the detrimental impacts of farming on river and coastal areas. BMPs for urban areas include storm water collection and treatment and the use of porous surfaces to reduce run-off and even out the storm hydrograph (Chapter 6). Hence, CZM can introduce management methods to protect the coast by tackling the problem both onshore and at source, thus controlling some of the issues raised in Chapter 6.

All aquatic life forms and water-based leisure activity ultimately depend on water quality for success. In the coastal zone, pressure on the water resource has traditionally been great, with a tendency to treat the seas as a large waste disposal unit. Discharges range from heated water to raw sewage, and include industrial pollutants, discharges through storm drains, and domestic drainage. All such discharges can have serious consequences for coastal habitats, as well as making the water unattractive for bathing. In addition to land-based sources, many industries actively dump at sea, both legally and illegally, on the principle of out of sight, out of mind. Often this dumped material is highly toxic or hazardous (Chapter 4).

Modern land-based sources are discharged well below the mean low-water mark of ordinary spring tides (MLWOST), and so become redistributed by tidal currents and do not recontaminate the land. They do, however, have the potential to affect marine organisms and fish stocks.

Many of the older sewage systems, and storm-water discharge outlets, do not discharge so far out, with the result that tidal currents can bring material back inshore, causing solid matter to be washed up on beaches, so that problems relating to odour and bacteria levels can result. Other problems with water-borne pollutants – that is, those which are water soluble – are not as readily detected, and so tend to go unnoticed until major environmental problems result. In contrast, the less ecologically serious problems of solid sewage on beaches tend to be noticed immediately, and treated by local authorities.

As part of the CZM principle, new outfalls have to be built to beyond MLWOST, in order to discharge out to sea in such a way that material is not brought back on tidal currents. For example, the discharge pipes for Weymouth and Portland in southern England were short, and many piped material directly into Weymouth Bay, with the result that much material washed ashore. A scheme to build a long outfall was proposed and put before a committee of people representing all the user groups of the area – that is, fishermen, conservationists, and the Royal Navy – and was amended accordingly to reduce any impacts or conflicts of interest. The result was a 2.5 km pipe discharging material directly into the English Channel. It was installed in a way that reduced individuals' fears, and reassured conservation groups that no environmental problems would result to Chesil Beach or adjacent habitats. New European legislation includes tighter controls on the quality of discharges, which means that much more land-based treatment of sewage and other discharges will need to occur before discharge into the sea.

When considering estuaries, the problems of disposal of dredged spoil are particularly acute because, since estuaries are often sites for harbour activity, these sediments can be highly contaminated, or occur in large quantities so as to need regular dredging. By adopting a wider coastal perspective, CZM can perceive wider uses for 'clean' material for the benefit of the coast, the alternative being to dump offshore in one of the licensed dump sites. Offshore disposal has been the traditional option for sewage or contaminated materials which are unsuitable for soft defence methods, because of the coasts involved. Alternative disposal practices include incineration (at fifteen times the cost) and processing (at between eight and ten times).

Leaving aside industrial uses, perhaps the prime uses of the coastal waters are for recreation and tourism. This can lead to some of the greatest conflicts which CZM plans have to deal with. They include trampling in sensitive areas and conflicts of usage, such as those caused by incompatible activities such as bird-watching and wild-fowling. To minimise such conflicts, the CZM plan needs to study the activities which occur along the coast, and to accommodate and zone usage so that each may occur without affecting others.

Biodiversity and habitat conservation

Coastal areas support a wide biodiversity, which can be severely reduced by loss of habitat. In England, English Nature has started to become aware of this problem, with its Campaign for a Living Coastline, and has identified loss in major coastal habitats and areas where re-creation is needed in order to maintain habitat levels (see Box 3.1).

In the USA habitat conservation plans (HCPs) are used to balance development and conservation interests. In this case, some states will actively seek areas to purchase for conservation and habitat creation in order to balance areas of land lost through development with the creation of replacement areas. This approach provides a good model for the English Nature scheme in the UK, where national areas of habitats are targeted to remain at those of the early 1990s.

Protecting access

As well as protecting the natural environment, management also needs to protect access to the shore. The coast is an immense recreational area, but as it becomes more developed the levels of access can be reduced, either by making areas inaccessible or by charging for use of facilities.

Protecting coastal development

Development along the coast can often lead to small towns and villages losing their identity when they are swallowed up in development. Many tourist areas are guilty of this form of identity loss, with coastal villages being developed so that they merge with other neighbouring villages undergoing similar development. As well as the loss of all identity and uniqueness, the result is a continuous development along much of a popular tourist coast, removing many possibilities for natural habitat management. Management planning needs to impose strict regulations on such areas, to preserve uniqueness. Canon Beach, in Oregon, has a ban on large fast-food outlets, to protect it from mass tourism and day trippers. Other ways include tight building regulations with regard to style and materials used, limits on how many floors a development can have, and tight rules concerning the provision of open spaces.

Construction of a management plan

It is important to bear in mind that a management plan aims to protect a stretch of coast or an estuary by allowing multi-usage by industry, members of the public, and developers. There are many potential areas of conflict between these groups and bad feeling if the whole thing is not done well. As the ultimate purpose is to allow citizens to enjoy the coast in a sustainable way, whether for recreation, work, or as a place to live, it is important that such people have a say in the eventual management plan.

Many of the coastal management plans which have been proposed have been developed by a panel of experts, including ecologists, sociologists, engineers, coastal scientists, etc. These plans act as a decree, with rules and regulations being issued from on high, and dictating to users of the environment what they can and cannot do. The net outcome of many of these plans is that they meet great opposition from users, who resent being told what they can and cannot do in their own neighbourhood, with the result that they can sometimes be completely disregarded and so fail to work.

In contrast, by far the most successful method is to work out a strategy for coastal management by incorporating as many views as possible, from as many users as possible. If the plan is based on the needs of the people whom it aims to serve, there is a greater chance that it will be successful. Clearly, there still needs to be some degree of overall control exerted by the experts constructing the plan, but overall, less resentment is felt if people are aware that everything possible is being done to protect their interests. The group responsible for plan construction needs to consult with all user groups, determining the type of use, frequency, and intensity, and also any ways in which improvements could be made. By building in as many interests as possible to a plan, it is possible to produce a better-received document, and one which is more likely to be accepted by the local population.

A second important consideration is that nothing is written in tablets of stone. The strategy or planning document is advisory, and subject to frequent updating and modification to allow for changes in usage patterns, relative impacts of users, and a changing environment. Any plan needs to be subject to review so as to incorporate areas of new activity, growth and decline of others, and changes in natural processes such as erosion and accretion.

Procedure

The first important decision is the exact form that the management is to take, and the area which it is expected to cover. This decision will govern the actions needed, and the authority which will be charged with its co-ordination and implementation. Regardless of the exact form of plan, it is important to remember that one of the main objectives is to improve the understanding of the coastal processes operating within the sediment cell, because only by doing this can effective management occur. An insight having been gained into how the coast is operating, other factors, such as predicting the likely future evolution of the coast, and the identification of all the assets within the plan area which are likely to be affected by such coastal change, can be provided for.

If all these are satisfactorily completed, then the final plan will be a document which will allow an assessment of strategic coastal defence options for the area of coast governed by the plan, and an agreed approach for individual sites, such as which sites will be defended, which will not, and which may be the location for managed realignments. The plan should also outline the future requirements for monitoring and measurement, and provide a reference source for planning authorities in order to allow continued assessment of how the plan is functioning, and whether it remains viable for that particular coastline. From a conservation point of view, the plan also needs to identify areas of important habitat, and areas of high ecological value, along with potential sites for improved habitat creation. The identification of possible sites for habitat creation can then be linked back to the coastal defence strategies because it is feasible to allow these to operate together, with new habitats being created as a form of coastal defence (Chapter 3). Finally, the plan should also facilitate the setting up of a mechanism which allows continued interaction between interested parties, and the continued updating of the strategic plan. Doing so causes the plan to remain a viable and workable document.

To achieve those objectives outlined above, the plan needs to incorporate information on four key areas. The key to any plan is a good understanding of how the coast operates. Only by knowing what the coast has done in the past can a prediction be made as to what is likely to happen in the future. This means that it is necessary to understand how the coast has changed over history, how it is operating at the present time, and, most importantly, how past and present behaviour relate in order to predict areas likely to need defences.

As defences already exist on many coasts, the plan needs to consider the likelihood of these continuing to be effective and the future need for upgrading. When the future of defences is being considered, the potential of removing coastal defence structures to 'free up' coastal sediment must be taken into account as a possible way of permitting the coast to function in as natural a state as possible.

Study of defences is relevant only in the context of what they are protecting. Not only should current defences be considered, but any future needs for flood or erosion protection should be determined, and areas where development should no longer be allowed should be highlighted. Coastal planning policies should normally provide only for those developments which require a coastal location. Ongoing dredging policies, restrictions on use, and protection of historic buildings also need to be considered.

Finally, the ecological and scientific importance of the whole natural coastal environment needs to be considered. Designation of statutory protection for areas of particular geomorphological, ecological, or geological interest needs to be highlighted. Plans will also need to take into account the fact of sea level rise, and the potential for habitat loss as a result of continued exploitation and coastal squeeze. Estimates of loss, such as those indicated for England by Pye and French (1993a), need to be built into any strategy plan to allow for areas to be set aside for the re-creation of habitat to combat any losses. The report should also designate areas for habitat re-creation, possibly in conjunction with managed retreat.

The production process

All the uses of the coastal area having been considered, relevant individuals and groups consulted, and the information required having been specified, it is paramount to obtain and present this information in a way which can be understood and appreciated by all who need to consult it. The production of a management plan can be split into two stages, data gathering and plan preparation.

By gathering as much data as possible covering the whole range of uses treated in the report, the plan can be produced to its fullest extent. For example, in the UK the Morecambe Bay strategy covers the following range of relevant issues: coastal defence, fisheries, heritage and landscape, industry, transport and development, land management, pollution, recreation, and tourism, and wildlife. Such categories can be used as the basis for many plans. Each of these areas forms the basis of data collection from as many sources as possible, including the environment itself, historical records, scientific literature, unpublished records, and members of the public/user groups/societies, etc.

Once the data have been obtained and interpreted, plan preparation has to occur. The section of coast to be managed can be divided into subunits for the sake of management, or treated as a whole, as in the case of Morecambe Bay. Within each of the categories used, it is necessary to make management decisions to cover the future of that particular category; that is, how will future infrastructure development be managed? How will recreation be controlled? What areas will be defended?

This document then becomes a draft plan, put out for consultation. Following this consultation period, amendments are made and the final report is produced. This report is not a statutory document, but is merely a reference and advisory document. Governments, however, make it a strong recommendation that any development is made with reference to such a strategy document.

Finally, the plan should specify a time when updating will take place. This is often after a period of five years, at which time all objectives and recommendations are reviewed, any new information assessed, and the report updated and reissued as a second version.

As a non-statutory document, the report can be considered only as advisory. Frequently, management reports concentrate on individual parts of the coastal cell, such as an urban frontage, embayment, or estuary. While the ultimate aim is to manage the coast on a coastal cell basis, it is important to use the management plans in conjunction with other plans covering different areas in the same coastal cell. Figure 7.1 details a framework which could be employed in order to do this.

Regions for coastal zone management

As we have already seen in Chapter 1, in order to manage the coastline to the best possible degree, it is necessary to take into account coastal processes. For example, we have seen that sediment movements around the coast occur in set patterns, with erosion from one area leading to deposition in another, and that such processes create a balance within clearly defined areas of the coast, or cells. It is important for CZM to take account of these cells, as any management schemes will have to take note of sediment movement, sources, and stores so as to avoid complications with the sediment budget. For example, if one authority were to protect a cliff from eroding, this would cut off a major source of sediment supply (see Chapter 3).

In summary, we are in a position to tackle many of the problems to which we have referred in earlier chapters. There is a workable framework in place which can be used as a basis for all management plans. By forming plans for all parts of the coast-line, particularly those which are heavily developed, we should be able to come up with a coast which is effectively managed with sensible development and the most suitable defence types, as well as being as natural as possible, permitting unimpeded movement of sediment and promoting wildlife habitats. Some people would argue that this is a rather ideal situation and could never be achieved. Certainly no one would ever suggest that all the world's coasts have to be naturalised, but it is clear that many parts can be. What is perhaps the most frustrating factor in all this is that

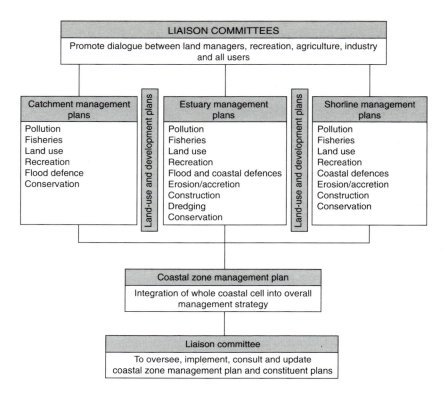

FIGURE 7.1 Framework for the integration of the various management plans into an overall coastal strategy

huge amounts of money are being spent on outdated forms of defence, when less money could be spent using newer ideas and methods to protect the coast more naturally, more effectively, and for longer before new investment is needed. The underlying problem lies in lack of education. Governments and national bodies need to inform and persuade their populations that to give land back to the sea is not such a bad thing, and in many cases can be a most efficient use of tax-payers' money.

The future of the coastline

Throughout this book, we have looked at various aspects of human impact along the coast associated with coastal protection and management. The various concepts can be divided up on the basis of coastal types, that is, strategies for wave-dominated, tide-dominated, and wind-dominated coasts, along with the concepts of hard *vis-à-vis* soft engineering, the role of human intervention, and the concepts of coastal zone management. We have also seen that CZM is being used increasingly to protect the whole coastal environment.

We have seen that the human impacts on coasts, and the application of coastal management, are in a state of change. Traditional methods are being replaced by new, more environmentally sensitive methods. We are seeing a shift from the traditional concrete barrier 'defend the coast at all costs – never allow the sea to erode the land' approach, to a softer, more habitat-based approach to defending our coasts. In the newer approach the coast functions in as natural a way as possible, resulting in a more natural coastline, interactive with the environment, while the same end results are still achieved.

This attitude change has three outcomes. First, the coastline becomes more environmentally friendly, has greater habitat potential, and has considerably more aesthetic appeal; second, being more natural, it can respond more easily to changes in environmental conditions (wind, waves, sea level), and so becomes more able to 'look after itself'; and third, the methods used are cheaper to implement.

We are entering an era of increased environmental concern for the coast, and increasing financial restrictions. In future, coasts will be protected by hard structures only where necessary. This means that some houses will be considered expendable, that actually removing defences and sacrificing land to the sea may be an environmentally better way of defence. These actions are likely to become increasingly common, but if they are to become more widely accepted, it is necessary to convince the general public of their viability, and also, perhaps more importantly, to compensate those whose houses and businesses are allowed to fall into the sea.

Implications of sea level rise for the future of the coast

The future of coastal management and defence also contains one other factor which we have not, as yet, mentioned in detail. So far we have talked about the coast only as a zone which is interacting with the sea. The other important factor which has to be built into our future predictions is what is actually happening to the sea itself. Of prime importance here is the fact that mean global sea level is increasing, causing the problem of sea level rise. The Intergovernmental Panel on Climate Change (1990) predicts the rates of this rise as being between 31 cm (low) and 110 cm (high) by 2100, with a best estimate of 66 cm.

This means that high tides become higher, wave base increases, and so the energy received at the coast may also increase (see Dolotov, 1992, for a detailed assessment). The prospect of the sea drowning large areas of land which are currently areas on which humans live is an emotive subject. Significantly, the problem is one which is common to low-lying areas the world over, irrespective of politics or wealth. Predictions of how the coast will react to such changes have been made possible by studying how the coast has changed in response to Holocene sea level rise, following the last glaciation.

Many countries already have systems in place to cope with sea level rise, such as sea defences and barrages. Many of these systems will need to be ungraded, or, more drastically, the land concerned allowed to be flooded by the sea with any population or land use having to relocate. Under conditions of accelerated sea level rise, coastal erosion may become more rapid, requiring either new defences, or methods of habitat creation, such as managed realignment. In some areas, such methods will not be a problem. In others, such as the Netherlands, the idea of allowing land to flood is a non-starter, as large areas of the country are already below sea level. Many countries now employ barrages to protect the larger cities; the Thames barrage in the UK has been constructed to withstand a 1,000 year flood, and so will be able to cope with sea level rise for some time.

The concept of building defences to protect our towns and cities is second nature to countries in the developed world that have the financial resources to cope, and we tend to take it for granted. However, the cost of such defence constructions may still be significant. Burby and Nelson (1991) quote data which indicate that the USA would need to defend an additional 1,025 km of its coast by 2100 following a 1.0 m rise in sea level. While this may represent a large financial burden, consider countries such as Bangladesh, which do not have the resources to construct hugely expensive coastal defences, and where predicted sea level rise is up to 345 cm by 2100 (range 56.2–345 cm; see Castro-Ortiz, 1994). Storm surges are a major cause of loss of life or of livelihood, and spoiling of land. Much of Bangladesh lies on the delta of the Ganges–Brahmaputra, hence high land refuges are not common. Elaborate engineering schemes have been proposed, but all are expensive and have serious environmental repercussions in their own right. A sea level rise of around 1 m by the middle of the twenty-first century would cause the loss of 23,000 km^2 of land. Some of Bangladesh's administrative districts would disappear altogether, as would 14 per cent of the area of crops and 29 per cent of its forest (Ince, 1990; Viles and Spencer, 1995). Much of the population, as a result, would need to move, although the possibility of new settlement within Bangladesh itself would be slight. The rise of 1 m is a mid-range prediction based on current forecasts for the area. What this represents in real terms is the loss of 14 million head of livestock, 100,000 households, 8,000 schools, 1,500 km of railways, 20,000 km of roads, 4,000 km^2 of mangroves. This would result in political and social upheaval on a scale rarely seen in the history of the world.

Bangladesh, however, is not alone in this problem. It is quite clear how rising seas can cause problems for society, as well as for the natural environment in general. Problems of habitat migration and habitat squeeze have been mentioned earlier. The effects on society of an increased frequency of flooding, population migration, and the loss of land and homes are also important considerations. Sea level rise is very much a global issue. While some countries are able to adapt, others are not. Ways in which this problem is having an effect on different countries become readily apparent when looking at individual case histories. A series of studies is presented below to illustrate the problems of some countries. A more extensive case study bibliography is given in Table 7.4.

The Maldives

There is considerable interest in the Maldives Islands in the Indian Ocean when considering problems of sea level rise, as may be expected in a nation whose highest

TABLE 7.4 Select bibliography of sea level rise studies from different countries

Country or area	Author(s)	Region
Argentina	Codignotto et al. (1992)	—
Australia	Belperio (1993)	Port Adelaide estuary
	Knighton et al. (1991)	Northern Australia
	Semeniuk (1994)	North-western Australia
	Wolanski and Chappell (1996)	Tropical Australian estuaries
	Woodroffe (1995)	Northern Australian mangroves
Bangladesh	Ali (1996)	—
	Castro-Ortiz (1994)	—
	Khalequzzaman (1994)	—
	Castro-Ortiz (1995)	—
	Tickell (1994)	—
Belgium	MacDonald and O'Connor (1996)	—
Canada	Orford et al. (1991)	Story Head, Nova Scotia
	Scott and Greenberg (1983)	Bay of Fundy
	Shaw and Forbes (1990)	Newfoundland
Caribbean	Bacon (1994)	—
Caspian Sea	Ignatov et al. (1993)	—
	Kaplin and Selivanov (1995)	—
China	Ren (1993)	—
	Yang and Zhu (1993)	Changjiang estuary
Denmark	Bruun (1993)	Skagen Spit, Jutland
	Jacobsen (1993)	Wadden Sea
Egypt	Frihy et al. (1996)	Alexandria
France	Day et al. (1995)	The Camargue
	MacDonald and O'Connor (1996)	—
Gambia	Jallow et al. (1996)	—
India	Ramanujam et al. (1995)	Gulf of Mannar
	Rao et al. (1990)	Nizampatnam Bay
	Shetye et al. (1990)	—
Italy	Bondesan et al. (1995)	North-east coast
	Pirazzoli (1991)	Venice
Netherlands	Bakker et al. (1993)	North Sea coasts
	den Elzen and Rotmans (1992)	North Sea coast
	Eitner (1996)	East Frisian barrier islands
	MacDonald and O'Connor (1996)	—
Philippines	Perez et al. (1996)	Manila Bay
South Africa	Hughes and Brundrit (1992)	—
	Hughes et al. (1991)	—
	Hughes et al. (1993)	Diep River, Cape Town
	Sowman et al. (1990)	—
Southern Hemisphere	Lutjeharms and Valentine (1991)	General countries

TABLE 7.4 continued

Country or area	Author(s)	Region
Spain	Sánchez–Arcillan *et al.* (1996)	Ebro delta
Tunisia	Oueslati (1992)	Gulf of Gaben
UK	Clayton (1990)	—
	French (1993)	North Norfolk
	French (1996a)	Severn estuary
	Hewlett and Birnie (1996)	Severn estuary
	Raffaelli (1992)	Scottish estuaries
	Tooley (1991)	—
USA	ASCE (1992)	—
	Canning (1991)	Washington State
	Daniels (1992)	South Carolina
	Day and Templet (1987)	Mississippi Delta
	Day *et al.* (1995)	Mississippi Delta
	Flanagan (1993)	Atlantic coast
	Fletcher (1992)	—
	Hackney and Cleary (1987)	North Carolina
	Klarin and Hershman (1990)	—
	Leatherman (1987)	Maryland
	London and Volonte (1991)	Myrtle Beach, South Carolina
	Maul and Martin (1993)	Key West, Florida
	Moorhead and Brinson (1995)	North Carolina
	Orson *et al.* (1985)	Atlantic coast/Gulf of Mexico
	Roemmich (1992)	South-west coasts
	Ross *et al.* (1994)	Florida Keys
	Wilcoxen (1986)	San Francisco
	Yon *et al.* (1996)	—

Note: A dash in the 'Region' column indicates that the work refers to the general coast of that country, rather than to any specific region.

point is just 3 m above present sea level. Regardless of the variation in current sea level rise prediction, even the lowest estimates would be sufficient to cause physical, social, and economic damage. The population is concentrated on twenty-five of the islands, although 275 out of the total 1,300 are inhabited to some extent. Virtually the whole of the population lives between 0.8 and 2 m above sea level (Ince, 1990; Pernetta, 1992; Pernetta and Sestini, 1989).

As with other low-lying tropical islands, the Maldives are able to generate their own living sea defences, in the form of coral atolls. Provided that the rate of sea level rise is not too rapid, the growth of corals should be able to keep pace with it and provide storm protection (though not protection from increased water levels). This is by no means a certainty, however, and just the increasing erosion potential of the sea may be sufficient to erode coral as fast as it grows. This situation will exist only for a finite period of time. If we assume that the growth of coral reefs can keep pace with rising sea levels, then some protection will be provided although the increased water levels will still be inundating the islands. At some point, whole islands may become submerged, leaving rings of coral around them.

The main areas of risk include the country's agriculture, its airport and hence tourism, and the areas of reclaimed land which have been important in coping with an increasing population. Problems of flooding are already real. Legislation is required to promote sensible development, although modification of buildings and houses to cope with sea level rise would require a great deal of the country's investment fund. What the problem does mean is that the Maldivian government has had significant experience with the problems of sea level rise, and is one of the few that have a history of taking the subject seriously. In reality, the Maldives can cope with only a small amount of slow sea level rise. Anything in excess of c. 1 m in total would cause serious problems for local infrastructure and population. There may well come a point, in this case, where defence is no longer a realistic option, or even possible, with island abandonment becoming the only viable alternative.

Australia and the Great Barrier Reef

The Great Barrier Reef runs along the east coast of Australia, but in this case the effect of sea level rise is not necessarily for the worst. Over the past 20,000 years or so, a period of considerable sea level change, the reef has responded to slight changes and kept pace with sea level. It could be argued that this has not been a gradual process, and that the reef has failed to keep pace but has regenerated itself once conditions become favourable again (Ince, 1990). Although this may appear an academic argument, it is important when considering the response of the reef to sea level rise (see also Stoddart, 1990). An ability to keep pace with rising sea level will ensure continued wave protection for the coast; the ability to catch up at a later date when conditions change, assuming that they will, could mean a loss of this protection, and increased coastal erosion problems.

Corals can grow at up to $8 \, mm \, yr^{-1}$, or continue to grow at depths of 10 m, a figure far in excess of even the most pessimistic of predictions. As a result, the most likely scenario with regard to the reef and sea level rise is not one of destruction, but one of radical change such as the regeneration of coral exposed on the surface. Sea level rise could, however, flood some of the coral islands, some of which are inhabited, and some of which contain rare flora and fauna.

In this case, CZM has to consider one further point. The uniqueness of these islands is in their flora and fauna. To interfere and protect will cause impacts on these species and may destroy the naturalness of their habitat. As a result, it is important to consider whether these ecosystems will retain their uniqueness if interfered with, or whether it may be better to leave alone and allow the effects of sea level rise to be felt by the islands concerned. If left alone, populations may adapt to change. If sea level rise were sufficient to drown an island, then the possibility of habitat creation elsewhere should be considered, such as designating another island as a sanctuary.

The Netherlands

Large parts of the Netherlands have been created through land claim from the sea, and half the total land mass is less than 5 m above sea level (Ince, 1990). The Dutch are considered to be in greater peril from rising seas than any other developed nation, although their history and awareness of the sea make them perhaps the best equipped to protect themselves from it.

The Dutch coast tends to be defended either by dunes or by dykes. Each will respond differently to sea level rise. The dyke system is considerable, and an order of magnitude or so greater than that of the UK. Each defence is built to withstand a flood with a return period of 10,000 years, which in general, means that they are built approximately 4–6 m higher than the highest predicted storm surge. The main impact of rising sea level is to increase the likelihood of occurrence of a 1 in 10,000 years flood to 1 in 2,000 years, although this would be partly offset by the increasing depth of the North Sea as levels increased (Ince, 1990).

The real problem is that one-third of the country lies below sea level, and a further 25 per cent is less than 1 m above it. As sea level rises, this land will need to be more heavily defended, and the drainage system upgraded to accommodate the increase in base level. Looking at the worst-case scenario for the area, a 1.5 m rise in sea level would involve a major change in the country's way of life. It would be necessary to close the mouth of the Rhine with sluices and locks, even though it provides access to Europe's biggest port (Rotterdam). Water pumping would be needed in large parts of the country, and saline water might penetrate drinking-water supplies, while dune systems might reactivate and migrate inland. To combat such problems, interference will almost certainly be the most viable option considering the degree of dependence on marginal land, and the infrastructure concerned. Blocking the Rhine might prevent flooding but would have a strong impact on fluvial processes and habitats a long way inland. The type of problem would be similar to that of a tidal barrage (Chapter 3), with increased sedimentation, retention of pollution, dredging requirements, increased water tables, and increased coastal erosion due to sediment starvation. Hence, as a result, it is possible that storm damage could be increased by the reduced coastal sedimentation, increasing wave base and, therefore, wave height and intensity.

An impact assessment using a modelling package (IMAGE) has been carried out to assess the problems of sea level rise in the Netherlands (Den Elzen and Rotmans, 1992). This has shown that the impacts are large and involve considerable cumulative costs for the next century. Action required includes increasing the height of dykes, prevention of dune erosion, amelioration of saline water intrusion into aquifers, and increased drainage.

Pacific islands

The islands of the Pacific are many and diverse. For food, many depend on fishing rather than agriculture, and many have already disappeared owing to rises in sea level already experienced (Middleton, 1995). In addition, many are low-lying coral atolls, which are also the most vulnerable to sea level rise. A study by the New Zealand Ministry of Foreign Affairs estimates that under the higher predictions, the entire populations of Kiribati, Tokelau, and Tuvalu, around 70,000 people, would need to be resettled as it is not feasible to adopt the sorts of measures of defence practised elsewhere. In essence, the worst-case scenario for sea level rise would result in the destruction of these countries and the forced relocation of their inhabitants (Ince, 1990).

Many of the other islands would also be seriously affected, with the relocation of capital cities, industry and commerce, and government being necessary. Other problems with agriculture and crop production would also arise. One of the major effects, however, would be caused by the modification of ocean currents due to ocean warming, especially in the context of fish population migration and the local economy, with the movement of migration pathways offshore and away from traditional feeding grounds (Hinrichsen, 1990; Ince, 1990).

Guyana

Guyana has an area the size of the UK., but a population of only around 900,000. Despite this small size, around 90 per cent of the population occupy just 3 per cent of the land area, much of this being located within a coastal zone of reclaimed land (Ince, 1990).

With much of the country's economy being based along the coastal agricultural land, a rise in sea level of 0.5 m would produce serious economic damage, and 1.5 m would force drainage measures to be undertaken in much of the coastal area. The solution is difficult, as only minor amounts of alternative agricultural land are available further inland. Alternative wealth generation may be possible from mineral resources, but this has never been verified. The only realistic proposition is to defend, which will require a huge financial and practical effort (Hinrichsen, 1990; Ince, 1990).

◆　　◆　　◆

It is clear that the problems of sea level rise are different for many countries, and it is for this reason that there can be no definite standards in CZM planning. Obviously, the problem of managing sea level rise is much more acute in some areas than in others, with the threat of island abandonment very real in some cases. Sea level rise is a component of CZM which is variable and not accurately predictable. When CZM strategies are being developed, there is always going to be this largely indefinite value, which will vary according to which predictive model is used. Sea level rise will continue to occur in the medium term irrespective of what happens to global emissions and global warming, resulting in increased problems of flooding, erosion, aquifer intrusion, habitat loss, alteration of wave climate and modifications to the position of the fresh water–salt water interface (Pernetta and Elder, 1992; Titus, 1986). What is needed as part of the CZM process is, along with national statutory requirements for CZM, national and international strategic plans to cope with sea level rise. This strategy must identify not only those sites where sea defences need to be upgraded, but also those areas where important fronting habitats could be lost, and identify a series of sites where new habitat can be created. In the extreme, this option must also consider abandonment to the sea.

If coastal protection and management are to occur in a sensible and holistic manner, we have to get away from the outdated view that we have to defend every-thing from the sea, and that no land should be eroded. What we see as the coastline today is just a snapshot in time – it is a transitory feature, constantly having bits added to it and bits taken off, and the only way of successfully managing the coast is to recognise this fact, and to incorporate it into a sound coastal strategy.

In contrast to the doom and gloom of the sea level rise scenarios discussed so far, there is an alternative argument which may partly offset some of the problems which we have discussed. Carter (1988) raises the point, as we have also suggested, that as sea levels increase, so the erosion rates along many coasts will increase owing to the increased water depth and wave activity. This increased erosion will increase the amount of sediment available for input into the coastal system. Hence, the increased sediment load will be transferred down-drift, resulting in beaches containing larger sediment volumes. It can be argued, therefore, that this will partly offset the problems of increased sea level rise, and may result in the stabilisation of beach profiles. This is an aspect of sea level rise which is difficult to predict, but we can draw comparisons from the post-glacial rises in sea level. What is more certain is that while increased protection may result, the build-up of beaches, as argued by Carter, will result in an increase in the abundance of sand-dominated coastlines, as opposed to those dominated by shingle or mud. Sandy beaches tend to be less steep and wider than shingle beaches, and so the degree of protection they will afford will be different from that of their muddy and shingle counterparts.

While this process may offer some hope, it does ignore the fact that for beaches to become stable under new sea level conditions, it is necessary for them to occupy a position in the tidal frame similar to the one they did at earlier sea level conditions. Hence, it will be necessary to allow landforms to migrate landwards or coastal squeeze will result. This process brings us back to the original problem of much of the coast, and a constant problem for CZM, which is that landward retreat cannot occur because of human development and the high value of land adjacent to the coast.

Prediction of future rise

The perception of sea level rise has developed from being an alarmist scientific speculation to becoming a more factually based prediction based on evidence readily available at first hand. However, predicting rise in sea level is notoriously difficult. Since the 1960s, models for prediction, and the data used, have improved, with the result that estimates of rise have been gradually revised downwards. Some recent estimates (Wigley and Raper, 1990) suggest a rate of 4–5 mm yr^{-1} until the year 2100.

Other work indicates only 1–2 mm yr^{-1}, representing a halving. What is clear, however, is that as more data become available, and models more refined, estimates will get more reliable, and that whatever the rate of sea level rise, that rate is critical in the formatting of CZM plans.

Predictions are always difficult to make, considering the general uncertainty of the natural environment, and the process of saying how the planet will respond to global warming, what policy will be adopted, and how these will be reflected in the processes which affect sea level rise is not a simple one. What is certain, however, is that rises in sea level are occurring, and will, in all probability, continue so long as global warming increases. Considering that today, after a rise of only 10–15 cm over the past century, many more parts of the world suffer from coastal flooding, and our attempts at flood prevention have only been partially successful, predictions of a further 13 cm in the next forty years only make the issue more pressing.

We have seen through examples the types of impacts which sea level rise will have on various coastlines. Although the problems are country specific, the basic components of the problem remain the same: the level of the sea is getting higher. The sheer size and speed of the sea level rise issue mean that there are many different viewpoints relating to the implications, most of which eventually point to policy as the only solution. Major questions relating to the problem of sea level rise include such aspects as how nations are supposed to balance the unknown costs of environmental damage caused by sea level rise, with solutions to the problem on which no costs have been determined. However, it does become evident that the greatest sufferers from sea level rise are not only some of the poorer nations, but also nations which have little or no responsibility for causing the problem in the first place.

Although the viewpoints are varied, the process of dealing with the problem is likely to be more easily stated. The places where the problem is being taken most seriously are the low-lying island states of the lesser-developed areas, the peoples with the most to lose as their actual existence is in the balance. Since most of these nations are already poor and have limited resource potential, coping with sea level rise on top of all this has been compared to playing a piano concerto while fighting off an unfed tiger. It is really a question of the more technologically advanced and wealthy nations meeting requests for financial and professional help in full, because after all, if we accept the anthropogenic cause of sea level rise as being global warming, responsibility can be traced directly to the industrialised nations.

The way forward

It is clear that the coastal zone faces a complex variety of human pressures which potentially have severe implications for the coastal environment. The key to grasping these issues and creating a rational strategy which fully addresses them is to treat the coastal zone as a whole, with the foreshore, hinterland, and offshore areas being parts of a single unit within which occur a complex series of interactions. We have seen from Table 7.3 how many countries have adopted this philosophy and have a single controlling authority for coastal management. We have also seen that in the UK, such things are not as well advanced. The problem with UK coastal management has been a lengthy process of lack of interest and problem realisation. A parliamentary select committee has now recognised that the current planning and management of the UK coastal zone is epitomised by the number of organisations with powers or responsibilities, and, as a result, no single body has the responsibility or authority to take an overview of the coastal zone, or even to ensure co-ordination. This situation leads to overlap, conflict, confusion, omission, piecemeal action, and lack of long-term planning and management, and, as a result, a poorly managed strategy.

The overlap between authorities poses many problems. As an example, the Dee estuary falls within the areas of two county councils and six district councils, and also falls partly within England and partly within Wales. This latter fact means that it also falls within the jurisdiction of both the Department of the Environment and the Welsh Office. Clearly, this situation verges on farce when one is trying to construct a management policy for this system.

Although there are many management plans in operation, none has any statutory regulatory authority, and not all are as comprehensive as they might be because of non-co-ordination between authorities. This is not to say, however, that plans do not have a role in statutory planning. Increasingly, many are being incorporated into development plans and official management policy for coastal areas. This formation of plans and co-ordination needs to be systematised to put it on a more clearly defined platform. In addition, the statutory planning powers of local authorities could be extended further out to sea to allow offshore activities to be included in coastal planning controls.

Things are improving to a significant extent. Local authorities are coming together as they realise the importance of the coastal zone, and of the adoption of an overall management strategy which could be beneficial for the coastal zone as a whole. The UK government, however, is still reviewing options and, as yet, has not taken the lead. It is essential that coastal management planes are co-ordinated under one body, whether as part of MAFF or as part of the DoE, because only in that way can plans be made uniform and of similar standards. At present, this task can only be monitored and overseen; there are no powers to intervene. The National Coasts and Estuaries Advisory Group is acting in this capacity, and has published reports dealing with good practice in management plan preparation (King and Bridge, 1994; National Coasts and Estuaries Advisory Group, 1993). However, as if emphasising the problems with too many interested parties, other reports on the same topics have also been produced by the DoE (Department of the Environment, 1995) and MAFF (Pethick and Burd, 1993; Ministry of Agriculture, Fisheries and Food et al., 1995; Ministry of Agriculture, Fisheries and Food and the Welsh Office, 1993).

In addition to this, English Nature has its own scheme, Campaign for a Living Coastline, which aims to maintain areas of coastal habitats and natural features at levels equivalent to their 1992 distribution and in a sustainable condition. This implies that while the preservation of human life is paramount, the protection of property must be balanced with the need to secure society's environmental assets for the future. Hence, this campaign is very much orientated towards soft engineering, with much emphasis being placed on freeing up the coast, rather than protecting it further.

This policy advocates the need for allowing coastal planning and management to utilise the power of natural forces to augment human efforts to protect the coast. This is significantly different from the historical approach of trying to stop natural processes and alter them, which only leads to conflict between humans and nature. Natural processes happen; humans cannot stop them, they can only alter them. The best way of protection is to get nature on your side, working with it rather than against it.

It is important to recognise that coastal areas will be lost, salt marshes will erode, dunes will migrate inland, shingle spits will breach and re-form. Where possible, this should be allowed to happen, with loss compensated for elsewhere. These actions are readily compatible with the idea of managed retreat.

The RSPB report *Save Our Shorebirds Campaign* (Royal Society for the Protection of Birds, 1993) recognises many of the features that are present in the Campaign for a Living Coastline. Its report goes slightly further in that it makes a series of recommendations to government on all aspects of planning. Many of these procedures have been accepted to varying degrees; many of the important ones, however, have not.

Summary of Chapter 7

This chapter has taken a slightly different aspect of the human intervention problem and considered coastal impacts as a management issue. We have seen that only by taking an holistic approach, by treating the coast as functional units, can successful management ever occur. It is possible, however, that in any developed part of the world some aspect of coastal usage will obstruct the ideal coastal plan, such as not being able to free up the sediment in a shingle bank because it has a nuclear power station on it. While there is a clear methodology outlining how the process is best achieved, coastal managers often have to accept a compromise to fit the management plan to what is actually present on the ground.

The most successful and effective coastal management schemes operate where the system of administration and control is simple. In situations of multi-control and competing interests there is bound to be a problem of non-communication. It is rather frustrating to have all the tools at hand to do the job, only to be prevented by authorities pulling in different directions. It is essential to sort out the logistics before success can be achieved at grassroots level. It is impossible to operate a successful coastal strategy if reference constantly has to be made to different groups with different controlling and governing interests. This remains an issue which needs sorting out in many countries.

The topic of coastal management also contains one of the largest unknowns, the problem of global sea level rise. We know from tide records that sea level is definitely rising. This may be considered a natural process of land movement relative to sea, but there are also other components caused by global warming, such as the melting of ice and the thermal expansion of sea water. These problems are not going to go away in the foreseeable future, and need major international consultation on emission controls before they are likely to do so.

The problem of sea level rise is a very real issue for coastal management. It is perhaps difficult for most people to appreciate that sea level rise means anything other than new defences, or a little more coastal erosion. For many nations, however, it is a matter of pure existence. Abandoning your island home is not something that many of us would wish to contemplate, but is an issue already facing people in the Maldives, for example. Where the problems are not as severe, it is perhaps better to adopt a management strategy that allows the coast to maintain its sediment inputs and stores as much as possible. Only by adopting these options can the long-term stability of the coast be best achieved.

There have been many important concepts and ideas introduced in this chapter. I have taken the issues raised in earlier chapters and grouped them into ways of tackling the problems, mainly by the process referred to as coastal zone management. The reader should leave this chapter with an understanding of:

1   the need for, and uses or, the coastal management process;
2   how the formation of coastal management plans is best achieved;
3   what a coastal plan contains and what it does;
4   the problems of sea level rise in management, and the impacts caused by it.

The final word

Throughout this text, we have covered many issues which relate to the broad topic of human impacts on coasts and estuaries. The intention of this book is that it should be considered not as a comprehensive guide to all impacts and their effects on coasts and estuaries, but more as a source for informing students of the sorts of problems, and their diversity, which face the coastal manager.

We can sum up coastal problems in two words: sediment and pollution. All the impacts considered in this book can relate to one of these two. All coastal instability inevitably comes down to either a lack of sediment or too much sediment. The human role in this is to bring about one of these by constructing defences to cut off sediment supply, or clear land to increase sediment yields. The presence of industry at the coast not only makes land more valuable, and therefore more in need of protection, thus cutting off sediment supply further, but also increases pollution, the second major area of impact.

In many respects, coastal management presents the ideal solution to these problems. The snag is, however, that many coastlines are already too developed to allow the ideal management plan to be put into effect. Encouragingly, however, there is an increasing awareness of the natural coast, and a recognition that this will function better in many situations than a concrete, artificial coast. We can look at this another way. The reason that sea walls are built is to stop the waves from hitting the vulnerable coast behind. The damage that waves do is a function of momentum and energy. If you are driving a car down a road, the car has energy and momentum, as does a wave. When you want to stop it, you apply a braking mechanism to reduce that momentum and energy gradually, until you come to a stop before hitting the car in front. When an aeroplane comes in to land, it also has energy and momentum, and so it increases its resistance to this by using its flaps, and also, once on the runway, it too employs a braking mechanism, thus reducing its energy and momentum before crashing into the terminal building. So then, considering that the problem of stopping anything is due to these two physical parameters, and that it is generally acknowledged that the best way of stopping and reducing these parameters is to employ some form of braking device, why, when we consider waves, do we typically allow them to run straight into concrete walls?

Intertidal areas can be considered the waves' braking mechanism, and so the use of these can reduce the need for sea walls. Reducing the need for sea walls will allow the coast to behave more naturally, and if it behaves more naturally, then the coast will be more capable of looking after itself. It is through the acceptance of this idea that soft engineering techniques are becoming increasingly common on coastlines. They are not the universal solution; hard engineering structures are still needed in some areas. The more areas which are managed by soft techniques, however, the better for the coastline, the better for conservation, and, in the end, the better for us.

References

Abdullah M.I. and Royle L.G. (1974) A study of the dissolved and particulate trace elements in the Bristol Channel. *Journal of the Marine Biological Association of the UK* 54: 581–597

Abouseida M.M. and Alsarawi M.A. (1990) Utilization and management of coastal areas in Kuwait. *Coastal Management* 18(4): 385–401

Adriaanse L.A. and Choosen J. (1991) Beach and dune nourishment and environmental aspects. *Coastal Engineering* 16: 129–146

Alam I.A. (1993) The 1991 Gulf war oil spill: lessons from the past and a warning for the future. *Marine Pollution Bulletin* 27: 357–360

Albakri D. (1996) A geomorphological approach to sustainable planning and management of the coastal zone of Kuwait. *Geomorphology* 17(4): 323–337

Alexander W.B., Southgate B.A., and Bassindale R. (1936) Summary of the Tees estuary investigation survey of the River Tees: Part II, The estuary, chemical and biological. *Journal of the Marine Biological Association of the UK* 20: 717–724

Ali A. (1996) Vulnerability of Bangladesh to climate change and sea level rise through tropical cyclones and storm surges. *Water Air and Soil Pollution* 92(1–2): 171–179

Allen J.R.L. (1987) Coal dust in the Severn Estuary, southwest U.K. *Marine Pollution Bulletin* 18: 169–174

—— (1988) Modern-period muddy sediments in the Severn Estuary (southwest U.K.): a pollution-based model for dating and correlation. *Sedimentary Geology* 58: 1–21

—— (1990) Late Flandrian shoreline oscillations in the Severn Estuary: change and reclamation at Arlingham, Gloucestershire. *Philosophical Transactions of the Royal Society of London* A330(1613): 315–334

—— (1991) Salt marsh accretion and sea-level movement in the inner Severn Estuary, southwest Britain: the archaeological and historical contribution. *Journal of the Geological Society* 148: 485–494

Allen J.R.L. and Rae J. (1986) Time sequence of metal pollution, Severn Estuary, southwestern U.K. *Marine Pollution Bulletin* 17: 427–431

Almadany I.M., Abdalla M.A., and Abdu A.S.E. (1991) Coastal zone management in Bahrain: an analysis of social, economic and environmental impacts of dredging and reclamation. *Journal of Environmental Management* 32(4): 335–348

Alsarawi M.A., Marmoush Y.R., Lo J.M., and Alsalem K.A. (1996) Coastal management of Failaka Island, Kuwait. *Journal of Environmental Management* 47(4): 299–310

Amir S. (1984) Israel's coastal program: resource protection through management of land-use. *Coastal Zone Management Journal* 12 (2–3): 189–223

Anders F.J. and Leatherman S.P. (1987) Effects of off-road vehicles on coastal fore-dunes at Fire Island, New York, USA. *Environmental Management* 11(1): 45–52

Anthony E.J. (1994) Natural and artificial shorelines of the French Riviera: an analysis of their interrelationship. *Journal of Coastal Research* 10(1): 48–58

Anthony E.J. and Cohen O. (1995) Nourishment solutions to the problem of beach erosion in France: the case of the French Riviera. In Healy M.G. and Doody J.P. (eds) *Directions in European Coastal Management*. Samara Publishing, Cardigan, pp. 199–206

Archer J.H. (1995) Bulgaria coastal management program: the World Bank funds development of the first Black Sea I.C.A.M. effort. *Ocean and Coastal Management* 26(1): 77–82

Archer J.H. and Knecht R.W. (1987) The United States national coastal zone management program: problems and opportunities in the next phase. *Coastal Management* 15(2): 103–120

ASCE (1992) Effects of sea level rise on bays and estuaries. *Journal of Hydraulic Engineering* 118(1): 1–10

Bacon P.R. (1994) Template for evaluation of impacts of sea level rise on Caribbean wetlands. *Ecological Engineering* 3(2): 171–186

Bacon S. and Carter D.J.T. (1991) Wave climate changes in the North Atlantic and North Sea. *International Journal of Climatology* 11: 545–588

Bagnold R.A. (1954) *The Physics of Wind-Blown Sand and Desert Dunes*. Methuen, London

Bakker J.P., de Leeuw J., Dijkema K.S., Leendertse P.C., Prins H.H.T., and Rozema J. (1993) Salt marshes along the coast of the Netherlands. *Hydrobiologia* 265(1–3): 73–95

Ballinger R.C, Smith H.D., and Warren L.M. (1994) The management of the coastal zone of Europe. *Ocean and Coastal Management* 22: 45–85

Barley G. (1993) Integrated coastal management: the Florida Keys example from an activist citizen's point of view. *Oceanus* 36(3): 15–18

Barrow C.J. (1995) *Developing the Environment: Problems and Management*. Longman, London

Barston R.P. (1994) International dimensions of coastal zone management. *Ocean and Coastal Management* 23: 93–116

Bashirullah A.K.M., Mahmood N., and Matin A.K.M. (1989) Aquaculture and coastal zone management in Bangladesh. *Coastal Management* 17(2): 119–127

Baye P. (1990) Ecological history of an artificial fore-dune ridge on a north–eastern barrier spit. In Davidson-Arnott R.G.D. (ed.) *Proceedings of the Symposium on Coastal Sand Dunes*. National Research Council of Canada, Ottawa, pp. 389–403

Beardall C.H., Dryden R.C., and Holzer T.J. (1991) *The Suffolk Estuaries: a Report by the Suffolk Wildlife Trust on the Wildlife and Conservation of the Suffolk Estuaries*. Suffolk Wildlife Trust, Saxmundham, Suffolk

Beatley T., Bower D.J., and Schwab A.K. (1994) *An Introduction to Coastal Zone Management.* Island Press, Washington, DC

Belperio A.P. (1993) Land subsidence and sea level rise in the Port Adelaide estuary: implications for monitoring the greenhouse effect. *Australian Journal of Earth Sciences* 40(4): 359–368

Bettinetti A., Pypaert P., and Sweerts, J.P. (1996) Application of an integrated management approach to the restoration project of the lagoon of Venice. *Journal of Environmental Management* 46(3): 207–227

Bird E.C.F. (1985) *Coastline Changes: A Global Review.* John Wiley, Chichester

—— (1987) The modern prevalence of beach erosion. *Marine Pollution Bulletin* 18: 151–157

—— (1990) Artificial beach nourishment on the shore of Port Philip Bay, Australia. *Journal of Coastal Research* 6: 55–68

—— (1993) *Submerging Coasts: The Effects of Rising Sea Level on Coastal Environments.* John Wiley, Chichester

Bocamazo L. (1991) Sea Bight to Manasques, New Jersey beach erosion coastal projects. *Shore and Beach* 59: 37–42

Bohlen W.F., Cundy D.F., and Tramontano J.M. (1979) Suspended material distribution in the wake of estuarine dredging operations. *Estuarine and Coastal Marine Science* 9(6): 699–711

Bondarenko V.S. (1990) Coastal management in the USSR: Perestroika on the coast. *Coastal Management* 18(4): 337–363

Bondesan M., Castiglioni G.B., Elmi C., Gabbianelli G., Marocco R., Pirazzoli P.A., and Tomasin A. (1995) Coastal areas at risk from storm surges and sea level rise in northeastern Italy. *Journal of Coastal Research* 11(4): 1354–1379

Boorman L.A. and Fuller R.M. (1977) Studies on the impact of paths on the dune vegetation at Winterton, Norfolk, England. *Biological Conservation* 12: 203–216

Boorman L.A. and Hazelden J. (1995) Saltmarsh creation and management for coastal defence. In Healy M.G. and Doody J.P. (eds) *Directions in European Coastal Management.* Samara Publishing, Cardigan

Bray M.J., Carter D.J., and Hooke J.M. (1992) *Sea Level Rise and Global Warming: Scenarios, Physical Impacts and Policies.* SCOPAC, Portsmouth

British Coal (1986) Cardiff and the coalfield. Unpublished British Coal report. British Coal Public Relations Office, Cardiff

Broadus J.M. and Gaines A.G. (1987) Coastal and marine area management in the Galápagos Islands. *Coastal Management* 15(1): 75–88

Brooke J.S. (1992) Coastal defence: the retreat option. *Journal of the Institute of Water and Environmental Management* 6: 151–157

Bruland K.W., Bertine K., Koide M., and Goldberg E.G. (1974) History of metal pollution in southern California coastal zone. *Environmental Science and Technology* 8: 425–432

Bruun P. (1993) Relation between growth of a marine foreland and sea level rise: The Skagen Spit, north Jutland, Denmark. *Journal of Coastal Research* 9(4): 1125–1128

Buchner G.E. (1979) Engineering aspects of strip land reclamation with special reference to the Wash. In Knights B. and Phillips A.J. (eds) *Estuarine and Coastal Land Reclamation and Storage.* EBSA, Saxon House, Farnborough

Buckley R. (ed.) (1992) *The Mediterranean Paradise under Pressure*. Understanding Global Issues no. 8. European School Book Publications, Cheltenham

Burbridge P.R. (1988) Coastal and marine resource management in the Strait of Malacca. *Ambio* 17(3): 170–177

Burby R.J. and Nelson A.C. (1991) Local government and public adaption to sea level rise. *Journal of Urban Planning and Development* 117(4): 140–153

Burd F. (1989) *The Saltmarsh Survey of Great Britain*. Research and Survey in Nature Conservation. Nature Conservancy Council, Peterborough

Campbell J.A., Whitelaw K., Riley J.P., Head P.G., and Jones P.D. (1988) Contrasting behaviour of dissolved and particulate nickel and zinc in a polluted estuary. *Science of the Total Environment* 71: 141–155

Campbell T.J. and Spadoni R.H. (1987) Beach restoration: an effective way to combat erosion at the southeast coast of Florida. *Shore and Beach* 50: 11–12

Canning D.J. (1991) Washington State sea level rise response project. *Northwest Environmental Journal* 7(2): 377–379

Carlson C.H. and Godfrey P.J. (1989) Human impact management in a coastal recreation and natural area. *Biological Conservation* 49: 141–156

Carter D.J.T. and Draper I. (1988) Has the North Atlantic become rougher? *Nature* 337: 494

Carter R.W.G. (1988) *Coastal Environments: An Introduction to the Physical, Ecological and Cultural Systems of Coastlines*. Academic Press, London

Carter R.W.G., Eastwood D.A., and Bradshaw P. (1992) Small-scale sediment removal from beaches in Northern Ireland: environmental impact, community perceptions, and conservation management. *Aquatic Conservation: Marine and Freshwater Ecosystems* 2: 95–113

Case R. (1984) *Coastal Management: A Case Study of Barton-on-Sea*. Geography 16–19 series, Longman, Harlow

Castilla J.C. (1996) The future Chilean marine park and preserves network and the concepts of conservation, preservation and management according to legislation. *Revista Chilena de Historia Natural* 69(2): 253–270

Castro-Ortiz C.A. (1994) Sea level rise and its impact on Bangladesh. *Ocean and Coastal Management* 23(3): 249–270

Charlier R.H. and Charlier C.C.P. (1995) Sustainable multiple-use and management of the coastal zone. *Environmental Management and Health* 6: 14–24

Charlier R.H. and de Meyer C.P. (1989) Coastal defence and beach renovation. *Ocean and Shoreline Management* 12: 525–543

—— (1992) Tourism and the coastal zone: the case of Belgium. *Ocean and Coastal Management* 18: 231–240

—— (1995a) Beach nourishment as efficient coastal protection. *Environmental Management and Health* 6: 26–34

—— (1995b) New developments in coastal protection along the Belgian coast. *Journal of Coastal Research* 11(4): 1287–1293

Chen D.G. (1989) Coastal zone development, utilization, legislation, and management in China. *Coastal Management* 17(1): 55–62

Chill J., Butcher C., and Dyson W. (1989) Beach nourishment with fine sediment at Carlsbad,

California. In American Society of Civil Engineers (ed.) *Coastal Zone '89*. ASCE, New York, pp. 2092–2103

Chou L.M. (1994) Marine environmental issues of southeast Asia: state and development. *Hydrobiologia* 285(1–3): 139–150

Christensen B. (1983) Mangroves: what are they worth? *Unasylva* 35: 2–15

Cicin-Sain B. (1993) Sustainable development and integrated coastal management. *Ocean and Coastal Management* 21: 11–43

Claridge G. (1994) Management of coastal ecosystems in eastern Sumatra: the case of Berbak wildlife reserve, Jambi Province. *Hydrobiologia* 285(1–3): 287–302

Clark R.B. (1989) *Marine Pollution*, 2nd edition. Clarendon Press, Oxford

—— (1992) *Marine Pollution*, 3rd edition. Oxford University Press, Oxford

Clayton K.M. (1989) The implications of climatic change. In Institution of Civil Engineers (ed.) *Coastal Management*. Thomas Telford Press, London, pp. 165–176

—— (1990) Sea level rise and coastal defences in the UK. *Quarterly Journal of Engineering Geology* 23(4): 283–287

—— (1993) *Coastal Processes and Coastal Management*. Countryside Commission, London

—— (1995) Predicting sea level rise and managing the consequences. In O'Riordan T. (ed.) *Environmental Science for Environmental Management* Longman, Harlow

Clements M. (1993) The Scarborough experience: Holbeck landslide, 3/4 June 1993. *Proceedings of the Institution of Civil Engineers – Municipal Engineer* 103: 63–70

Codignotto J.O., Kokot R.R., and Marcomini, S.C. (1992) Neotectonism and sea level changes in the coastal zone of Argentina. *Journal of Coastal Research* 8(1): 125–133

Cooper N. (1996) Beach replenishment: implications for source and longevity from results of the Bournemouth schemes. Unpublished conference abstract, Coastal Defence and Nature Conservation, Portsmouth University, March

Cornforth R. (1994) Integrated coastal zone management: the Pacific way. *Marine Pollution Bulletin* 29: 10–13

Correia F., Dias J.A., Bosk T., and Ferraira O. (1996) The retreat of the eastern Quarteira cliffed coast (Portugal), and its possible causes. In Jones P.S., Healy M.G., and Williams A.T. (eds) *Studies in European Coastal Management*. Samara Publishing, Cardigan

Cortright R. (1987) Fore-dune management on a developed shoreline: Nedonna Beach, Oregon. In American Society of Civil Engineers (ed.) *Coastal Zone '87* New York, pp. 1343–1356

Coughlin J. (1979) Aspects of reclamation in Southampton Water. In Knights B. and Phillips A.J. (eds) *Estuarine and Coastal Land Reclamation and Storage*. EBSA, Saxon House, Farnborough

Craik W. (1996) The Great Barrier Reef marine park, Australia: a model for regional management. *Natural Areas Journal* 16(4): 344–353

Crawford D. (1992) The injured coastline: a parliamentary report on coastal protection in Australia. *Coastal Management* 20(2): 189–198

Crecelius E.A. (1975) The geochemical cycle of arsenic in Lake Washington and its relation to other elements. *Limnology and Oceanography* 20: 441–451

Cullen P. (1980) Management of the Australian coastal zone. *Marine Pollution Bulletin* 11(12): 342–343

—— (1982) Coastal zone management in Australia. *Coastal Zone Management Journal* 10(3): 183–212

Daily Telegraph (1996) Oil spillages around Europe since 1967. 'Electronic Telegraph', 21 February

Daniels R.C. (1992) Sea level rise on the South Carolina coast: two case studies for 2000. *Journal of Coastal Research* 8(1): 56–70

Davidson N.C., d'A Laffoley D., Doody J.P., Gordon J., Key R., Drake C.M., Pienkowski M.W., Mitchell R., and Duff K.L. (1991) *Nature Conservation and Estuaries in Great Britain.* Nature Conservancy Council, Peterborough

Davis R.A. (1996) *Coasts.* Prentice-Hall, Englewood Cliffs, NJ

Davison A.T., Nicholls R.J., and Leatherman S.P. (1992) Beach nourishment as a coastal management tool: an annotated bibliography on the developments associated with artificial nourishment of beaches. *Journal of Coastal Research* 8(4): 984–1022

Day J.C. and Gamble D.B. (1990) Coastal zone management in British Columbia: an institutional comparison with Washington, Oregon and California. *Coastal Management* 18(2): 115–141

Day J.W. and Templet P.H. (1989) Consequences of sea level rise: implications from the Mississippi Delta. *Coastal Management* 17(3): 241–257

Day J.W., Pont D., Hensel P.F., and Ibanez C. (1995) Impacts of sea level rise on deltas in the Gulf of Mexico and the Mediterranean: the importance of pulsing events to sustainability. *Estuaries* 18(4): 636–647

de Jong D.J. and de Jong Z. (1995) The consequences of a one year tidal reduction of 35% for the saltmarshes of the Oosterscheldt (south-west Netherlands). *Coastal Zone Topics: Process, Ecology and Management* 1: 41–50

de Lange W.P. and Healy T.R. (1990) Renourishment of a flood-tidal delta adjacent beach, Taurange Harbour, New Zealand. *Journal of Coastal Research* 6(3): 627–640

de Leeuw J., de Munck W., Apon L.P., Herman P.M.J., and Beeftink W.G. (1995) Changes in saltmarsh vegetation following the construction of the Oosterscheldt storm surge barrier. *Coastal Zone Topics: Process, Ecology and Management* 1: 35–40

de Ronde J.G. (1993) What will happen to the Netherlands if sea level rise accelerates? In Warrick R.A., Barrow E.M., and Wigley T.M.L. (eds) *Climate and Sea Level Change.* Cambridge University Press, Cambridge, pp. 322–335

de Ruig J.H.M. and Louisse C.J. (1991) Sand budget trends along the Holland coast. *Journal of Coastal Research* 7(4): 1013–1026

de Souza C.M.M., Pestana M.H.D., and Lacardes L.D. (1986) Geochemical partitioning of heavy metals in sediments of three estuaries along the coast of Rio de Janeiro, Brazil. *Science of the Total Environment* 58: 195–198

Degong C. (1989) Coastal zone development, utilization, legislation, and management in China. *Coastal Management* 17: 55–62

Delderfield E.R. (1981) *The Lynmouth Flood Disaster*, 10th edition, 4th reprint. ERD Publications, Exeter

den Elzen M.G.J. and Rotmans J. (1992) The socio-economic impact of sea level rise on the Netherlands: a study of possible scenarios. *Climatic Change* 20(3): 169–195

Department of the Environment (1992) *Coastal Zone Protection and Planning.* HMSO, London

—— (1995) *Policy Guidelines for the Coast.* HMSO, London

Dixon K. and Pilkey O.H. (1991) Summary of beach replenishment on the US Gulf of Mexico shoreline. *Journal of Coastal Research* 7: 249–256

Dolan R. and Davis R.E. (1992) An intensity scale for Atlantic coast north–east storms. *Journal of Coastal Research* 8: 840–853

Dolotov Y.S. (1992) Possible types of coastal evolution associated with the expected rise in sea level caused by the greenhouse effect. *Journal of Coastal Research* 8(3): 719–726

Doody J.P. (1992) Sea defence and nature conservation: threat or opportunity? *Aquatic Conservation: Marine and Freshwater Ecosystems* 2: 275–283

—— (in prep.) *Coastal Habitat Loss: An Historical Review of Man's Impact on the Coastline of Great Britain*. Joint Nature Conservation Committee, Peterborough

Drury E. (1995) Revision of the European Union bathing water directive. In Healy M.G. and Doody J.P. (eds) *Directions in European Coastal Management*. Samara Press, Cardigan, pp. 107–112

Dyrynda P. (1996) Barrages within estuaries: ecological lessons from the Tawe Development. *Marine Update no. 25*. WWF-UK, Godalming

Earll B. (1994) UK government publishes its views on coastal zone management. *Marine Pollution Bulletin* 28(1): 5–6

Eid E.M.E. and Fawzi M.A. (1991) Egyptian approach towards appropriate use of coastal zones on the Red Sea. *Marine Pollution Bulletin* 23: 331–337

Eitner V. (1996) Geomorphological response of the Frisian barrier islands to sea level rise: an investigation of past and future evolution. *Geomorphology* 15(1): 57–65

Eitner V. and Ragutzki G. (1994) Effects of artificial beach nourishment on nearshore sediment distribution (Island of Norderney, southern North Sea). *Journal of Coastal Research* 10(3): 637–650

Elderfield H. and Hepworth A. (1975) Diagenesis, metals, and pollution in estuaries. *Marine Pollution Bulletin* 6: 85–87

English Nature (1993) *Strategy for the Sustainable Use of England's Estuaries*. English Nature, Peterborough

Erlenkeuser M., Suess M., and Willkomm E. (1974) Industrialisation affects heavy metal and carbon isotope concentrations in recent Baltic Sea sediments. *Geochimica et Cosmochimica Acta* 38: 823–842

Finkl C.W. (1996) What might happen to America's shorelines if artificial beach replenishment is curtailed: a prognosis for southeastern Florida and other sandy regions along regressive coasts. *Journal of Coastal Research* 12(1): iii–ix

Finney B. and Huh C. (1989) High resolution of sediment records of heavy metals from Santa Monica and San Pedro Basins, California. *Marine Pollution Bulletin* 20: 181–187

Fischer D.W. (1985) Shoreline erosion: a management framework. *Journal of Shoreline Management* 1: 37–50

Fischer D.W., Stone G.W, Morgan J.P., and Henningsen D.E. (1986) Integrated multidisciplinary information for coastal management, Florida. *Journal of Coastal Research* 2(4): 437–447

Flanagan R. (1993) Beaches on the brink. *Earth*, November: 24–33

Fletcher C.H. (1992) Sea level trends and physical consequences: application to the U.S. shore. *Earth Science Reviews* 33: 73–109

Flinn M.W. (1986) *The History of the British Coal Industry*: Volume 2, *1700–1830: The Industrial Revolution*. Clarendon Press, Oxford

Flood S., Cocklin C., and Parnell K. (1993) Coastal resource management conflicts and community action at Mangawhai, New Zealand. *Coastal Management* 21(2): 91–111

Fluharty D.L. (1984) The chrysanthemum and the coast: management of coastal areas in Japan. *Coastal Zone Management Journal* 12(1): 1–17

Förstner U. (1989) *Contaminated Sediments*. Lecture Notes in the Earth Sciences no. 21. Springer-Verlag, Berlin

Förstner U. and Müller G. (1973) Heavy metal associations in river sediments: a response to environmental pollution. *Geoforum* 14: 53–62

Foster G.A., Healy T.R., and de Lange W.P. (1996) Presaging beach renourishment from a nearshore dredge dump mound, Mt Maunganui Beach, New Zealand. *Journal of Coastal Research* 12(2): 395–405

Fox J. (1997) Has the Sidmouth sea defence scheme been successful in maintaining beach levels, restricting sediment movement and protecting the town from storm action? Unpublished BSc dissertation, Department of Geography, University of Lancaster

Fraser R.J. (1993) Removing contaminated sediments from the coastal environment: the New Bedford Harbour project example. *Coastal Management* 21: 155–162

French J.R. (1993) Numerical simulation of vertical marsh growth and adjustment to accelerated sea level rise, north Norfolk, UK. *Earth Surface Processes and Landforms* 18(1): 63–81

French P.W. (1990) Coal-dust: a marker pollutant in the Severn Estuary and Bristol Channel. Unpublished PhD thesis, PRIS, University of Reading

—— (1991) Natural set-back at Pagham Harbour. Unpublished report

—— (1993a) Areal distribution of selected pollutants in contemporary intertidal sediments of the Severn Estuary and Bristol Channel. *Marine Pollution Bulletin* 26: 692–697

—— (1993b) Post-industrial pollution levels in contemporary Severn Estuary intertidal sediments compared to pre-industrial levels. *Marine Pollution Bulletin* 26: 30–35

—— (1996a) Implications of a salt marsh chronology for the Severn Estuary based on independent lines of dating evidence. *Marine Geology* 135(1–4): 115–125

—— (1996b) Long-term temporal variability of copper, lead, and zinc in salt marsh sediments of the Severn Estuary, U.K. *Mangroves and Saltmarshes* 1(1): 59–68

Frihy O.E. (1996) Some proposals for coastal management of the Nile delta coast. *Ocean and Coastal Management* 30(1): 43–59

Frihy O.E., Dewidar K.M., and Elraey M.M. (1996) Evaluation of coastal problems at Alexandria, Egypt. *Ocean and Coastal Management* 30(2–3): 281–295

Fuller R.M. and Randall R.E. (1988) The Orford shingles, Suffolk, UK: classic conflicts in coastline management. *Biological Conservation* 46(2): 95–114

Gabrielides G.P., Golik A., Loizides L., Marino M.G., Bingel F., and Torregrossa M.V. (1991) Man-made garbage pollution on the Mediterranean coastline. *Marine Pollution Bulletin* 23: 437–441

Georgison J.P. and Day J.C. (1995) Port administration and coastal zone management in Vancouver, British Columbia: a comparison with Seattle, Washington. *Coastal Management* 23(4): 265–291

Gerges M.A. (1993) On the impacts of the 1991 Gulf war on the environment of the region: general observations. *Marine Pollution Bulletin* 27: 305–314

Giardino J.R., Bednarz R.S., and Bryant J.T. (1987) Nourishment of San Luis Beach, Texas: an assessment of impact. In American Society of Civil Engineers (ed.) *Coastal Sediments '87*, vol. 2. ASCE, New York, pp. 1145–1157

Glynn P.W. (1994) State of coral reefs in the Galápagos Islands: natural vs anthropogenic impacts. *Marine Pollution Bulletin* 29(1–3): 131–140

Goldberg E.D. (1994) *Coastal Zone Space: Prelude to Conflict?* UNESCO, Paris

Gomez E.D., Aliño P.M., Yap H.T., and Licuanan W.Y. (1994) A review of the status of Philippine reefs. *Marine Pollution Bulletin* 29(1–3): 62–68

Goodhead T. and Johnson D. (1996) *Coastal Recreation and Management: The Sustainable Development of Maritime Leisure*. E. and F.N. Spon Publishing, London

Goodwin P. and Williams P.B. (1992) Restoring coastal wetlands: the California experience. *Journal of the Institute of Water and Environmental Management* 6: 709–719

Goudie A. (1981) *The Human Impact: Man's Role in Environmental Change*. Blackwell, Oxford

Granja H.M. (1996) Some examples of inappropriate coastal management practice in north-west Portugal. In Jones P.S., Healy M.G., and Williams A.T. (eds) *Studies in European Coastal Management*. Samara Publishing, Cardigan, pp. 121–128

Grenell P. (1991) Non-regulatory approaches to management of coastal resources and development in San Francisco Bay. *Marine Pollution Bulletin* 23: 503–507

Gribben I. (1984) The world's beaches are vanishing. *New Scientist*, 10 May

Grieve H. (1959) *The Great Tide*. Essex County Council, Chelmsford

Griggs G.B. and Fulton-Bennett K.W. (1987) Failure of coastal protection at Seacliff State Beach, Santa Cruz county, California, USA. *Environmental Management* 11(2): 175–182

Grip K. (1992) Coastal and marine management in Sweden. *Ocean and Coastal Management* 18(2–4): 241–248

Guilcher A. and Hallégouët B. (1991) Coastal dunes in Brittany and their management. *Journal of Coastal Research* 7(2): 517–533

Gupta R.S., Fondekar S.P., and Alagarsamy R. (1993) State of oil pollution in the northern Arabian Sea after the 1991 Gulf oil spill. *Marine Pollution Bulletin* 27: 85–91

Guy W.E. (1983) Florida's coastal management program: a critical analysis. *Coastal Zone Management Journal* 11(3): 219–248

Guzmán H.M. (1991) Restoration of coral reefs in Pacific Costa Rica. *Conservation Biology* 5(2): 189–195

Hackney C.T. and Cleary W.J. (1987) Salt marsh loss in south-eastern North Carolina lagoons: impacts of sea level rise and inlet dredging. *Journal of Coastal Research* 3(1): 93–97

Hale L.Z. and Olsen S.B. (1993) Coral reef management in Thailand. *Oceanus* 36(3): 27–34

Hall M.J. and Pilkey O.H. (1991) Effects of hard stabilisation on dry beach width for New Jersey. *Journal of Coastal Research* 7(3): 771–785

Hallégouët B. and Guilcher A. (1990) Moulin Blanc artificial beach, Brest, western Brittany, France. *Journal of Coastal Research* Special Issue 6: 17–20

Hamilton E.I. and Clarke K.C. (1985) The recent sediment history of the Esk Estuary, Cumbria, U.K.: the application of radiochronology. *Science of the Total Environment* 35: 325–386

Hamilton E.I., Watson P.G., Cleary J.J., and Clifton R.J. (1979) The geochemistry of the recent sediments of the Bristol Channel and Severn Estuary. *Marine Geology* 31: 139–182

Hamilton-Taylor J. (1979) Enrichments of zinc, lead and copper in recent sediments of Windermere, England. *Environmental Science and Technology* 13: 693–697

Hanley J.R. and Couriel D. (1992) Coastal management issues in the Northern Territory: an assessment of current and future problems. *Marine Pollution Bulletin* 25(5–8): 134–142

Hannel F.G. (1955) Climate. In MacInnes C.M. and Whittard W.F. (eds) *Bristol and Its Adjoining Counties*. University of Bristol, Bristol, pp. 47–66

Hansom J.D. (1988) *Coasts*. Cambridge University Press, Cambridge

Hanson H. and Lindh G. (1993) Coastal erosion: an escalating threat. *Ambio* 22: 188–195

Hanson H., Jönsson L., and Broms B. (1984) *Beach Erosion in Liberia: Causes and Remedial Measures*. Report 3090, Dept. of Water Resource Engineering, University of Lund, Sweden

Harrison P. and Parkes J.G.M. (1983) Coastal zone management in Canada. *Coastal Zone Management Journal* 11(1–2): 1–11

Harvey N. (1988) Coastal management issues for the mouth of the River Murray, South Australia. *Coastal Management* 16(2): 139–149

Haward M. (1993) Ocean and coastal zone management in Australia and New Zealand: current initiatives. *Ocean and Coastal Management* 19(3): 292–299

—— (1995) Institutional design and policy making down under: developments in Australian and New Zealand coastal management. *Ocean and Coastal Management* 26(2): 87–117

Healy M.G. and Doody J.P. (eds) (1995) *Directions in European Coastal Management*. Samara Press, Cardigan

Herlihy A.J. (1982) *Coast Protection Survey*. Report, appendix and maps in two volumes. Department of the Environment, London

Hewlett R. and Birnie J. (1996) Holocene environmental change in the inner Severn estuary, UK: an example of the response of estuarine sedimentation to relative sea level change. *The Holocene* 6(1): 49–61

Hickman T. and Cocklin C. (1992) Attitudes towards recreation and tourism development in the coastal zone: a New Zealand study. *Coastal Management* 20(3): 269–289

Hildreth R.G. (1992) Australian coastal management: some North-American perspectives on recent Queensland and other initiatives. *Coastal Management* 20(3): 255–268

Hinrichsen D. (1990) *Our Common Seas: Coasts in Crisis*. Earthscan, London

Hobbs A.J. and Shennan I. (1980) Remote sensing of saltmarsh reclamation in the Wash, England. *J. Shoreline Management* 2: 181–198

Holman R.A. and Bowen A.J. (1982) Bars, bumps and holes: models for the generation of complex beach topography. *Journal of Geophysical Research* 87: 457–468

Holmes P.R. (1988) Tolo Harbour: the case for integrated water quality management in a coastal environment. *Journal of the Institute of Water and Environmental Management* 2(2): 171–179

Hong S.Y. (1991) Assessment of coastal zone issues in the Republic of Korea. *Coastal Management* 19(4): 391–415

Hong S.Y. and Lee J. (1995) National level implementation of Chapter 17: the Korean example. *Ocean and Coastal Management* 29(1–3): 231–249

Hoozemans F.M.J. and Wiersma J. (1992) Is mean wave height in the North Sea increasing? *Hydrographic Journal* 63: 13–15

Horrill J.C., Darwall W.R.T., and Ngoile M. (1996) Development of a marine protected area: Mafia Island, Tanzania. *Ambio* 25(1): 50–57

Houston J.A. and Jones C.R. (1987) The Sefton coast management scheme: project and process. *Coastal Management* 15(4): 267–297

Hudson B.J. (1996) *Cities on the Shore: The Urban/Littoral Frontier*. Pinter, London

Hughes P. and Brundrit G.B. (1992) An index to assess South African vulnerability to sea level rise. *South African Journal of Science* 88(6): 308–311

Hughes P., Brundrit G.B., and Shillington F.A. (1991) South African sea level measurements in the global context of sea level rise. *South African Journal of Science* 87(9): 447–453

Hughes P., Brundrit G.B., Swart D.H., and Bartels A. (1993) The possible impacts of sea level rise on the Diep River/Rietvel system, Cape Town. *South African Journal of Science* 89(10): 488–493

Hunt A., Jones J., and Oldfield F. (1984) Magnetic measurements and heavy metals in atmosphere particulates of anthropogenic origin. *Science of the Total Environment* 33: 129–139

Hutchings P., Payri C., and Gabrie C. (1994) The current status of coral reef management in French Polynesia. *Marine Pollution Bulletin* 29(1–3): 26–33

Hydraulics Research Ltd (1986 *et seq.*) *A Macro-review of the Coastline of England and Wales.* 8 volumes. Hydraulics Research, Wallingford.

Ignatov Y.I., Kaplin P.A., Lukyannova S.A., and Solovieva G.D. (1993) Evolution of the Caspian Sea coasts under conditions of sea level rise: model for coastal change under increasing greenhouse effect. *Journal of Coastal Research* 9(1): 104–111

Imperial M.T. and Hennessey T.M. (1996) An ecosystem-based approach to managing estuaries: an assessment of the National Estuary Program. *Coastal Management* 24(2): 115–139

Ince M. (1990) *The Rising Seas*. Earthscan, London

Inman D. and Nordström C. (1971) On the tectonic and morphological classification of coasts. *Journal of Geology* 79: 1–21

Intergovernmental Panel on Climate Change (1990) *Climate Change: The IPCC Scientific Assessment*. Cambridge University Press, Cambridge

Iriberri O.C.A. (1984) Integrated approach to coastal zone management in Puerto Galera, Oriental Mindero, Philippines. *Water Science and Technology* 16(3–4): 433–440

Itoh K., Chikuma M., and Tanaka H. (1987) Levels of selenium and metals in sediment cores as determined by ^{210}Pb dating techniques. *Bulletin of Environmental Contamination and Toxicology* 39: 214–223

Jacobsen N.K. (1993) Shoreline development and sea level rise in the Danish Wadden Sea. *Journal of Coastal Research* 9(3): 721–729

Jallow B.P., Barrow M.K.A., and Leatherman S.P. (1996) Vulnerability of the coastal zone of the Gambia to sea level rise and development of response strategies and adaptation options. *Climate Research* 6(2): 165–177

Job D. (1993) Coastal management: Start Bay, Devon. *Geography Review*, November: 13–17

Johnson D.W. (1919) *Shore Processes and Shoreline Development*. John Wiley, New York

Jokiel P.L., Hunter C.L., Taguchi S., and Watarai L. (1993) Ecological impact of a fresh water 'reef-kill' in Kanehoe Bay, Oahu, Hawaii. *Coral Reefs* 12: 177–184

Joliffe I.P. (1981) *An Investigation into Coastal Erosion Problems in the Northern Part of the Isle of Man: Causes, Effects, and Remedial Strategies.* Isle of Man Harbour Board, Douglas

Jones J.R., Cameron B., and Fisher J.J. (1993) Analysis of cliff retreat and shoreline erosion: Thompson Island, Massachusetts, USA. *Journal of Coastal Research* 9(1): 87–96

Jones R.E. (1978) Heavy metals in the estuarine environment. *Water Research Centre Technical Report Number TR73*

Juhasz F. (1991) An international comparison of sustainable coastal zone management policies. *Marine Pollution Bulletin* 23: 595–602

Kàdomatsu T., Uda T., and Fujiwara K. (1991) Beach nourishment and field observations of beach changes on the Toban coast facing Seto Inland Sea. *Marine Pollution Bulletin* 23: 155–159

Kalicharran S. and Diab R. (1993) Proposals for rehabilitation and management of Isipingo Lagoon and estuary, South Africa. *Environmental Management* 17(6): 759–764

Kaplin P.A. and Selivanov A.O. (1995) Recent coastal evolution of the Caspian Sea as a natural model for coastal responses to the possible acceleration of global sea level rise. *Marine Geology* 124(1–4): 161–175

Katz A. (1989) Coastal resource management in Belize: potentials and problems. *Ambio* 18(2): 139–141

Kelletat D. (1992) Coastal erosion and protection measures at the German North Sea coast. *Journal of Coastal Research* 8(3): 699–711

Kenchington R. (1993) Tourism in coastal and marine environments: a recreational perspective. *Ocean and Coastal Management* 19: 1–16

Khalequzzaman M.D. (1994) Recent floods in Bangladesh: possible causes and solutions. *Natural Hazards* 9(1–2): 65–80

King C.A.M. (1959) *Beaches and Coasts.* Edward Arnold, London

King G. and Bridge L. (1994) *Directory of Coastal Planning and Management Initiatives in England.* National Coasts and Estuaries Advisory Group, Maidstone

King S.E. and Lester J.N. (1995) The value of saltmarsh as a sea defence. *Marine Pollution Bulletin* 30: 180–189

Kirby R. (1990) The sediment budget of the erosional intertidal zone of the Medway Estuary, Kent. *Proceedings of the Geological Association* 101: 63–77

Klarin P. and Hershman M. (1990) Response of coastal zone management programs to sea level rise in the United States. *Coastal Management* 18(2): 143–165

Knecht R.W. and Archer J. (1993) Integration in the US coastal zone management program. *Ocean and Coastal Management* 21: 183–199

Knecht R.W., Cicinsain B., and Fisk G.W. (1996) Perceptions of the performance of state coastal zone management programs in the United States. *Coastal Management* 24(2): 141–163

Knighton A.D., Mills K., and Woodroffe C.D. (1991) Tidal creek extension and salt water intrusion in northern Australia. *Geology* 19(8): 831–834

Koekebakker P. and Peet G. (1987) Coastal zone planning and management in the Netherlands. *Coastal Management* 15(2): 121–133

Koike K. (1990) Artificial beach construction on the shores of Tokyo Bay, Japan. *Journal of Coastal Research* Special Issue 6: 45–54

Krumgalz B.S., Fainstein G., Gorfunkel L., and Nathan Y. (1990) Fluorite in recent sediments as a trap of trace metal contaminants in an estuarine environment. *Estuarine, Coastal and Shelf Science* 30: 1–15

Lamb H.H. (1982) *Climate, History and the Modern World*. Methuen, London

Leafe R. (1992) Northey Island: an experiment in set-back. *Earth Science Conservation* 31: 21–22

Leatherman S.P. (1979) *Barrier Island Handbook*. National Park Service Cooperative Research Unit, University of Massachusetts

—— (1987) Beach and shoreface response to sea level rise: Ocean City, Maryland, USA. *Progress in Oceanography* 18(1–4): 139–149

Leatherman S.P. and Godfrey P.J. (1979) *The Impact of Off-Road Vehicles on Coastal Ecosystems in Cape Cod National Seashore*. National Park Service Cooperative Research Unit, University of Massachusetts

Lee E.M. (1993) The political ecology of coastal planning and management in England and Wales: policy responses to the implications of sea level rise. *Geographical Journal* 159: 169–178

Lelliott R.E.L. (1989) Evolution of the Bournemouth defences. In Institution of Civil Engineers (ed.) *Coastal Management*. Thomas Telford Press, London, pp. 263–277

Liddle M.J. (1973) The effect of trampling and vehicles on natural vegetation. Unpublished PhD thesis, University College of North Wales

Lincolnshire County Council (1980) Reclamation on the Lincolnshire coast: a discussion. Unpublished discussion document, LCC

Lindén O. (1990) Human impact on tropical coastal zones. *Nature and Resources* 26: 3–11

Livesey J. (1997) The Morecambe coast: an analysis of sedimentation around the breakwaters and implications for management of the coast. Unpublished BSc dissertation, Department of Geography, University of Lancaster

London J.B. and Volonte C.R. (1991) Land-use implications of sea level rise: a case study of Myrtle Beach, South Carolina. *Coastal Management* 19(2): 205–218

Louisse C.J. and Kuik T.I. (1990) Coastal defence alternatives in the Netherlands. In Louisse C.I., Stive M.I.F., and Wiesma H.I. (eds) *The Dutch Coast: Report on a Session of the 22nd International Conference on Coastal Engineering*. Delft Hydraulics

Louisse C.J. and van der Meulen F. (1991) Future coastal defence in the Netherlands: strategies for protection and sustainable development. *Journal of Coastal Research* 7(4): 1027–1041

Lu K.Y. (1990) Marine and coastal management in China: a planning approach. *Coastal Management* 18(4): 365–384

Lutjeharms J.R.E. and Valentine H.R. (1991) Sea-level changes: consequences for the Southern Hemisphere. *Climatic Change* 18: 317–337

Ly C.K. (1980) The role of the Akosombo Dam on the Volta river in causing coastal erosion in central and eastern Ghana (west Africa). *Marine Geology* 37: 323–332

McAtee J.W. and Drawe D.L. (1980) Human impact on beach and foredune vegetation of North Padre Island, Texas. *Environmental Management* 4(6): 527–538

McCaffrey R.J. and Thompson J. (1980) A record of the accumulation of sediment and trace metals in a Connecticut salt marsh. *Advances in Geophysics* 22: 165–236

McClusky D.S. (1989) *The Estuarine Ecosystem*. Blackie, Glasgow

MacDonald N.J. and O'Connor B.A. (1996) Changes in wave impact on the Flemish coast due to increased mean sea level. *Journal of Marine Systems* 7(2–4): 133–144

McFarland S., Whitcombe L., and Collins M. (1994) Recent shingle beach renourishment schemes in the U.K.: some preliminary observations. *Ocean and Coastal Management* 25: 143–149

Mallik T.K. (1987) Coastal zone management program in Kerala, India. *Environmental Geology and Water Sciences* 10(2): 95–102

Maragos J.E. (1993) Impacts of coastal construction on coral reefs in the US affiliated Pacific Islands. *Coastal Management* 21: 235–269

Maragos J.E., Evans C.W., and Holthus P.F. (1985) Reef corals in Kanehoe Bay six years before and after termination of sewage discharges. *Proceedings of the 5th International Coral Reef Congress* 4: 189–194

Maul G.A. and Martin D.M. (1993) Sea level rise at Key West, Florida 1846–1992: America's longest instrument record. *Geophysical Research Letters* 20(18): 1955–1958

May V.J. (1969) Reclamation and shoreline changes in Poole Harbour, Dorset. *Dorset Natural History and Archaeological Society Proceedings* 90: 141–154

Medway Ports Authority (1975) *The River Medway: Historical and general account 1911*. Unpublished report, redrawn and reprinted in 1975. Medway Ports Authority, Sheerness

Meinesz A., Lefevre J.R. and Astier J.M. (1991) Impact of coastal development on the infra-littoral zone along the south-east Mediterranean shore of continental France. *Marine Pollution Bulletin* 23: 343–347

Merschrod K. (1989) In search of a strategy for coastal zone management in the third world: notes from Ecuador. *Coastal Management* 17(1): 63–74

Meyer I. (1991) Coastal protection: the race against time. *Geography Review* 4: 10–12

Middleton N. (1995) *The Global Casino: An Introduction to Environmental Issues*. Edward Arnold, London

Middleton R. and Grant A. (1990) Heavy metals in the Humber Estuary: Scrobicularia clay as a pre-industrial datum. *Proceedings of the Yorkshire Geological Association* 48: 75–80

Miles G., Fuavo V., and Smith A. (1995) Implementing Agenda 21: ocean, coasts and the Barbados outcomes in the Pacific region. *Ocean and Coastal Management* 29(1–3): 125–138

Miller G.T. (1994) *Living in the environment*, 8th edition. International Thompson Publishing, Belmont, CA

Miller, M.L. (1993) The rise of coastal and marine tourism. *Ocean and Coastal Management* 20: 181–199

Ministry of Agriculture, Fisheries and Food and The Welsh Office (1993) *Strategy for Flood and Coastal Defence in England and Wales*. MAFF/WO, London

Ministry of Agriculture, Fisheries and Food and The Welsh Office, Association of District Councils, English Nature and National Rivers Authority (1995) *Shoreline Management Plans: A Guide for Coastal Defence Authorities*. MAFF, London

Moller J.T. (1990) Artificial beach nourishment on the Danish North Sea coast. *Journal of Coastal Research* Special Issue 6: 1–9

—— (1992) Balanced coastal protection on a Danish North Sea coast. *Journal of Coastal Research* 8(3): 712–718

Montoya F.J. (1991) An administrative regulation pattern of coastal management for Mediterranean Sea: Spanish Shores Act (July 1988). *Marine Pollution Bulletin* 23: 769–771

Moorhead K.K. and Brinson M.M. (1995) Response of wetlands to rising sea level in the lower plain of North Carolina. *Ecological Applications* 5(1): 261–271

Morgan R., Jones T.C., and Williams A.T. (1993) Opinions and perceptions of England and Wales heritage coast beach users: some management implications from the Glamorgan Heritage Coast, Wales. *Journal of Coastal Research* 9(4): 1083–1093

Müller G. (1981) Heavy metals and other pollutants in the environment: a chronology based on the analysis of dated sediments. In *Heavy Metals in the Environment*. Amsterdam, pp. 12–17

Mumby P.J., Raines P.S., Gray D.A., and Gibson J.P. (1995) Geographic information systems: a tool for integrated coastal zone management in Belize. *Coastal Management* 23(2): 111–121

National Coasts and Estuaries Advisory Group (1993) *Coastal planning and management*. NCEAG, Maidstone

Nayak B.U, Chandramohan P. and Desai B.N. (1992) Planning and management of the coastal zone in India: a perspective. *Coastal Management* 20(4): 365–375

Newman D.E. (1976) Beach replenishment: sea defences and a review of the artificial beach replenishment. *Proceedings of the Institution of Civil Engineers* 1(60): 445–460

Ngoile M.N., Lindén O., and Coughanowr C.A. (1995) Coastal zone management in eastern Africa including the Island States: a review of issues and initiatives. *Ambio* 24(7–8): 448–457

Nordstrom K.F. (1988) Dune grading along the Oregon coast USA: a changing environmental policy. *Applied Geography* 8: 101–116

—— (1994) Beaches and dunes of human altered coasts. *Progress in Physical Geography* 18(4): 497–516

O'Brien R.J. (1988) Western Australia non-statutory approach to coastal zone management: an evaluation. *Coastal Management* 16(3): 201–214

Oldfield F. and Scoullos M. (1984) Particulate pollution monitoring in the Elefis Gulf: the role of mineral magnetic studies. *Marine Pollution Bulletin* 15: 229–231

Olsen S.B. (1987) A collaborative effort in developing the integrated coastal resources management for Ecuador. *Coastal Management* 15(1): 97–101

Orford J.D. (1988) Alternative interpretation of man-induced shoreline changes in Rosslare Bay, southeast Ireland. *Transactions of the Institute of British Geographers* 13: 65–78

Orford J.D. and Carter R.W.G. (1985) Storm generated rock armouring on a sand–gravel ridge barrier system, southeastern Ireland. *Sedimentary Geology* 42(1–2): 65–82

Orford J.D. and Forbes D.L. (1991) Gravel barrier migration and sea level rise: some observations from Story Head, Nova Scotia, Canada. *Journal of Coastal Research* 7(2): 477–489

Organisation for Economic Co-operation and Development (1993) *Coastal Zone Management: Integrated Policies*. OECD, Paris

Orlova G. and Zenkovich V.P. (1974) Erosion on the shores of the Nile Delta. *Geoforum* 18: 68–72

Orson R., Panageotou W., and Leatherman S.P. (1985) Response of tidal salt marshes of the U.S. Atlantic and Gulf coasts to rising sea levels. *Journal of Coastal Research* 1(1): 29–37

Oueslati A. (1992) Salt marshes in the Gulf of Gabes (southeastern Tunisia): their morphology and recent dynamics. *Journal of Coastal Research* 8(3): 727–733

Owens M. (1984) Severn Estuary: an appraisal of water quality. *Marine Pollution Bulletin* 15: 41–47

Ozhan E. (1996) Coastal zone management in Turkey. *Ocean and Coastal Management* 30(2–3): 153–176

Pacyna J.M. and Münch J. (1987) Atmospheric emissions of As, Cd, Pb, and Zn from industrial sources in Europe. In Lindberg S.E. and Hutchinson T.C. (eds) *Heavy Metals in the Environment*, volume 1. New Orleans, pp. 20–25

Palanques A., Plana F., and Maldonado A. (1990) Recent influence of man on the Ebro margin sedimentation system, northwestern Mediterranean Sea. *Marine Geology* 95: 247–263

Parker, R. (1980) *Men of Dunwich: The Story of a Vanished Town.* Paladin Grafton Books, London

Parkinson M. (1980) The salt marshes of the Exe estuary. *Reports and Transactions of the Devonshire Association for the Advancement of Science* 112: 17–41

—— (1985) The Axe estuary and its marshes. *Reports and Transactions of the Devonshire Association for the Advancement of Science* 117: 19–62

Paskoff R. (1987) L'érosion des plages. *La Recherche* 14: 20–28

Pearce F. (1993) When the tide comes in . . . *New Scientist*, 2 January

—— (1996) Black tide engulfs marine reserves. *New Scientist*, 2 March

Pearce R.J. (1991) Management of the marine environment in Western Australia: an ecosystem approach. *Marine Pollution Bulletin* 23: 567–572

Perez R.T., Feir R.B., Carandang E., and Gonzalez E.B. (1996) Potential impacts of sea level rise on the coastal resources of Manila Bay: a preliminary vulnerability assessment. *Water Air and Soil Pollution* 92(1–2): 137–147

Pernetta J.C. (1992) Impacts of climate change and sea level rise on small island states: national and international response. *Global Environmental Change* 2: 19–31

Pernetta J.C. and Elder D.L. (1992) Climate, sea level rise and the coastal zone: management planning for global changes. *Ocean and Coastal Management* 18: 113–160

Pernetta J.C. and Sestini G. (1989) The Maldives and the impacts of expected climate change. *UNEP Regional Seas Reports and Studies 104.* UNEP, Nairobi

Pethick J. (1981) Long-term accretion rates on tidal saltmarshes. *Journal of Sedimentary Petrology* 51: 571–577

—— (1984) *An Introduction to Coastal Geomorphology.* Edward Arnold, London

—— (1992) Natural change. In Institution of Civil Engineers (ed.) *Coastal Zone Planning and Management.* Thomas Telford Press, London, pp. 49–64

Pethick J. and Burd F. (1993) Coastal defence and the environment: a guide to good practice. MAFF, London

Pickering K.T. and Owen L.A. (1994) *An Introduction to Global Environmental Issues.* Routledge, London

Pirazzoli P.A. (1991) Possible defences against a sea level rise in the Venice area, Italy. *Journal of Coastal Research* 7(1): 231–248

Plant N.G. and Griggs G.B. (1992) Interactions between nearshore processes and beach morphology near a sea wall. *Journal of Coastal Research* 8(1): 183–200

Prestage M. (1990) Appeasing the gods. *Observer*, 21 October

Price A.R.G. (1993) The Gulf: human impacts and management initiatives. *Marine Pollution Bulletin* 27: 17–27

Price A.R.G., Wrathall T.J., Medley P.A.H., and Al-Moamen A.H. (1993) Broadscale changes in coastal ecosystems of the western Gulf following the 1991 Gulf war. *Marine Pollution Bulletin* 27: 143–147

Price B. (1986) Pollution in the Severn Estuary. Unpublished report to Greenpeace

Pringle A.W. (1987) Physical processes shaping the intertidal and subtidal zones. In Robinson N.A. and Pringle A.W. (eds) *Morecambe Bay: An Assessment of Present Ecological Knowledge.* Morecambe Bay Study Group and Centre for North West Regional Studies, University of Lancaster

Psuty N.P., Eugenia M., and Moreira S.A. (1992) Characteristics and longevity of beach nourishment at Praia da Rocha, Portugal. *Journal of Coastal Research* 8(3): 660–676

Psuty R. (1983) Coastal management programs on the eastern seaboard of the USA. *Journal of the Geological Society* 140: 971

Pye K. (1991) Saltmarshes on the barrier coastline of north Norfolk, eastern England. In Allen J.R.L. and Pye K. (eds) *Saltmarshes: Morphodynamics, Conservation and Engineering Problems.* Cambridge University Press, Cambridge

Pye K. and French P.W. (1993a) *Targets for Coastal Habitat Recreation.* Research and Survey in Nature Conservation no. 13. English Nature, Peterborough

—— (1993b) *Erosion and accretion processes on British salt marshes,* volumes 1–5. Final report to MAFF. Cambridge Environmental Research Consultants Ltd, Cambridge

Quélennec R.E. (1987) Coastal erosion in west and central Africa: an outlook on natural and man-made causes and consequences for the protection and management of coastal areas. *Nature and Resources* 23

Raffaelli D. (1992) Conservation in Scottish estuaries. *Proceedings of the Royal Society of Edinburgh B – Biological Sciences* 100: 55–76

Rajasuriya A., Ranjith M.W., de Silva N., and Öhman M.C. (1995) Coral reefs of Sri Lanka: human disturbance and management issues. *Ambio* 24(7–8): 428–437

Ramanujam N., Mukesh M.V., Sabeen H.M., and Preeja N.B. (1995) Morphological variations in some islands in the Gulf of Mannar, India. *Journal of the Geological Society of India* 45(6): 703–708

Ranwell D.S. and Boar R. (1986) *Dune Management Guide.* Institute of Terrestrial Ecology, Natural Environment Research Council, Huntingdon

Rao P.S., Rao G.K., Rao N.V.N.D., and Swamy A.S.R. (1990) Sedimentation and sea level variations in Nizampatnam Bay, east coast of India. *Indian Journal of Marine Sciences* 19(4): 261–264

Ratas U. and Purmann E. (1995) Human impact on the landscape of small islands in the west-Estonian archipelago. *Journal of Coastal Conservation* 1: 119–126

Reed D.J. (1992) Effect of weirs on sediment deposition in Louisiana coastal marshes. *Environmental Management* 16(1): 55–65

Ren M.E. (1993) Relative sea level changes in China over the last 80 years. *Journal of Coastal Research* 9(1): 229–241

Richmond R.H. (1993) Coral reefs: present problems and future concerns resulting from anthropogenic disturbance. *American Zoologist* 33: 524–536

Riddell K.J. and Young S.W. (1992) The management and creation of beaches for coastal defence. *Journal of the Institute of Water and Environmental Management* 6(5): 588–597

Riggs P. (1994) Fisheries and coastal management in the Republic of Estonia. *Coastal Management* 22(1): 25–48

Rijkswaterstaat (1990) *A New Coastal Defence Policy for the Netherlands*

Roberts A.G. and McGown A. (1987) A coastal area management system as developed for Seasalter–Reculver, north Kent. *Proceedings of the Institution of Civil Engineers: Design and Construction* 82: 777–797

Roemmich D. (1992) Ocean warming and sea level rise along the southwest United States coast. *Science* 257(5068): 373–375

Ross M.S., O'Brien J.J. and Sternberg L.D.L. (1994) Sea level rise and reduction in pine forests in the Florida Keys. *Ecological Applications* 4(1): 144–156

Rowan J.S., Higgitt D.L., and Walling D.E. (1993) Inclusion of Chernobyl-derived radiocaesium into reservoir sediment sequences. In McManus J. and Duck R.W. (eds) *Geomorphology and Sedimentology of Lakes and Reservoirs*. John Wiley, Chichester

Royal Society for the Protection of Birds (1993) *A Shore Future: RSPB Vision for the Coast*. The RSPB Save our Shorebirds Campaign. RSPB, Sandy, Bedfordshire

Ryan P.G. and Swanepoel D. (1996) Cleaning beaches: sweeping the rubbish under the carpet. *South African Journal of Science* 92: 275–276

Salomons W. and Förstner U. (1984) *Metals in the Hydrocycle*. Springer-Verlag, Berlin

Sánchez-Arcillan A., Jimenez J.A., Stive M.J.F., Ibanez C., Pratt N., Day J.W., and Capobianco M. (1996) Impacts of sea level rise on the Ebro delta: a first approach. *Ocean and Coastal Management* 30(2–3): 197–216

Sayre W.O. and Komar P.D. (1988) The Jump-Off Joe landslide at Newport, Oregon: history of erosion, development and destruction. *Shore and Beach* 56: 15–22

Schwartz M.G., Juanes J., Foyo J., and Garcia G. (1991) Artificial nourishment at Varadero Beach, Cuba. In American Society of Civil Engineers (ed.) *Coastal Sediments '91*. ASCE, New York, pp. 2081–2088

Scott D.B. and Greenberg D.A. (1983) Relative sea level rise and tidal development in the Fundy tidal system. *Canadian Journal of Earth Sciences* 20(10): 1554–1564

Semeniuk V. (1994) Predicting the effect of sea level rise on mangroves in northwestern Australia. *Journal of Coastal Research* 10(4): 1050–1076

Seys J., Meire P., Loosen J., and Craeymeersch J. (1995) Long-term changes in the intertidal macrobenthos fauna at eight permanent sites in the Oosterscheldt: effects of the construction of the storm surge barrier – preliminary results. In Jones N.V. (ed.) *Coastal Zone Topics: Process, Ecology and Management* 1: 51–60

Shapiro H.A. (1984) Coastal area management in Japan: an overview. *Coastal Zone Management Journal* 12(1): 19–56

Shaw J. and Forbes D.L. (1990) Relative sea level change and coastal response, northeast Newfoundland. *Journal of Coastal Research* 6(3): 641–660

Shetye S.R., Gouveia A.D., and Pathak M.C. (1990) Vulnerability of the Indian coastal region to damage from sea level rise. *Current Science* 59(3): 152–156

Sidaway R. (1995) Recreation and tourism on the coast: managing impacts and resolving conflicts. In Healy M.G. and Doody J.P. (eds) *Directions in European Coastal Management*. Samara Publishing, Cardigan, pp. 71–78

Sidaway R. and van der Voet J.Z.M. (1993) *Getting on Speaking Terms: Resolving Conflicts between Recreation and Conservation in Coastal Zone Areas of the Netherlands.* Centre for Recreation and Tourism Studies, University of Wageningen, the Netherlands

Singer C. *et al.* (1956) *A History of Technology,* volume 2. Oxford University Press, Oxford

Sloan N.A. and Sugandhy A. (1994) An overview of Indonesian coastal environmental management. *Coastal Management* 22(3): 215–233

Smith R.A. (1992) Conflicting trends of beach resort development: a Malaysian case. *Coastal Management* 20: 167–187

Smith S.V., Kimmerer W.J., Laws E.A., Brock R.E., and Walsh T.W. (1981) Kanehoe Bay sewage diversion experiment: perspectives on ecosystem responses to natural purtibation. *Pacific Science* 35: 1–42

Soegiarto A. (1981) Research and training programs on coastal resources management in Indonesia. *Bulletin of Marine Science* 31(3): 811

Sorensen J. (1990) An assessment of Costa-Rica coastal management program. *Coastal Management* 18(1): 37–63

Sorensen J. and Brandani A. (1987) An overview of coastal management efforts in Latin America. *Coastal Management* 15(1): 1–25

Sowman M.R. (1993) The status of coastal zone management in South Africa. *Coastal Management* 21(3): 163–184

Sowman M.R., Glazewski J.I., Fuggle R.F., and Barbour A.H. (1990) Planning and legal responses to sea level rise in South Africa. *South African Journal of Science* 86(7–10): 294–298

Stanley D.J. and Warne A.G. (1993) Nile Delta: recent geological evolution and human impact. *Science* 260: 628–634

Steers J.A. (1946) *The Coastline of England and Wales.* Cambridge University Press, Cambridge

Stoddart D.R. (1990) Coral reefs and islands and predicted sea level rise. *Progress in Physical Geography* 14(4): 147–171

Suárez de Vivero J.L. (1992) The Spanish Shores Act and its implications for regional coastal management. *Ocean and Coastal Management* 18: 307–317

Tabucanon M.S. (1991) State of coastal resource management strategy in Thailand. *Marine Pollution Bulletin* 23: 579–586

Taylor D. (1979) The effect of discharges from three industrial estuaries on the distribution of heavy metals in coastal sediments of the North Sea. *Estuarine and Coastal Marine Science* 8(4): 387–393

Templet P.H. (1986) American Samoa: establishing a coastal area management model for developing countries. *Coastal Zone Management Journal* 13(3–4): 241–264

Tickell O. (1995) Impending disaster for Bangladesh. *Geographical Magazine*, May

Tilmans M.K., Baarse G., van Pagee A., and Fontana D. (1985) Coastal zone management in Emilia-Romagna, Italy. In Schröder P.C. (ed.) *Sea Level Rise: A Selective Retrospection.* Delft Hydraulics, the Netherlands

Titus J.G. (1986) Greenhouse effect, sea level rise and coastal zone management. *Coastal Zone Management Journal* 14(3): 147–171

—— (1991) Greenhouse effect and coastal wetland policy: how America could abandon an area the size of Massachusetts at minimum cost. *Environmental Management* 15(1): 39–58

Tooley M.J. (1991) Future sea level rise – A working paper – Discussion. *Proceedings of the Institution of Civil Engineers Part 1 – Design and Construction* 90: 1101–1105

Tooley M.J. and Jelgersma S. (1992) *Impacts of Sea Level Rise on European Coastal Lowlands.* IBG Special Publication no. 27. London

Towle E.L. (1985) St Lucia: Rodney Bay/Gros Inlet. In Geoghagan T. (ed.) *Proceedings of the Caribbean Seminar on Environmental Impact Assessment.* Institute Resource and Environmental Studies. Dalhousie University, Halifax, Nova Scotia

Turner K.M. (1995) Managed realignment as a coastal management option. *Biological Journal of the Linnean Society* 56: 217–219

Valentin H. (1952) *Die Küsten der Erde.* Petermanns Geografische Mitteilungen Ergänzungshaft 246. Justies Perthes, Gotha

van Dijk H.W.J. (1994) Integrated management and conservation of Dutch coastal areas: premises, practice and solutions. *Marine Pollution Bulletin* 29(6–12): 609–616

van der Meulen F. and Salman A.H.P.M. (1996) Management of Mediterranean coastal dunes. *Ocean and Coastal Management* 30(2–3): 177–195

Verhagen H.J. (1990) Coastal protection and dune management in the Netherlands. *Journal of Coastal Research* 6(1): 169–179

—— (1996) Analysis of beach nourishment schemes. *Journal of Coastal Research* 12(1): 179–185

Viles H. and Spencer T. (1995) *Coastal Problems: Geomorphology, Ecology, and Society at the Coast.* Edward Arnold, London

Walker F. (1972) *The Bristol Region.* Thomas Nelson, London

Walsh J.P. (1982) The coastal zone management program: will revenue sharing save it? *Oceanus* 25(4): 20–28

Walton T.L. and Purpura J.S. (1977) Beach nourishment along the southeastern Atlantic and Gulf coasts. *Shore and Beach* 45: 10–18

Warren L.J. (1981) Contamination of sediments by lead, zinc and cadmium: a review. *Environmental Pollution* B2: 401–436

Watt J., Woodhouse T., and Jones D.A. (1993) Intertidal clean-up activities and natural regeneration on the Gulf coast of Saudi Arabia from 1991 to 1992 after the 1991 Gulf oil spill. *Marine Pollution Bulletin* 27: 325–331

Weggell J.R. and Sorenson R.M. (1991) Performance of the 1986 Atlanta City, New Jersey beach nourishment project. *Shore and Beach* 59: 29–36

Wigley T.M.L. and Raper S.C.B. (1990) Future changes in global mean temperature and thermal expansion-related sea level rise. In Warrick R.A. and Wigley T.M.L. (eds) *Climate and Sea Level Change: Observations, Projections and Implications.* Cambridge University Press, Cambridge, pp. 111–133.

Wilcoxen P.J. (1986) Coastal erosion and sea level rise: implications for Ocean Beach and San Francisco West Side transport project. *Coastal Zone Management Journal* 14(3): 173–191

Willmington R.H. (1983) The nourishment of Bournemouth beaches 1974–1975. In Institution of Civil Engineers (ed.) *Shoreline Protection.* Thomas Telford Press, London, pp. 157–162

Wolanski E. and Chappell J. (1996) The response of tropical Australian estuaries to a sea level rise. *Journal of Marine Systems* 7(2–4): 267–279

Woodroffe C.D. (1995) Response of tide-dominated mangrove shorelines in northern Australia to anticipated sea level rise. *Earth Surface Processes and Landforms* 20(1): 65–85

Woodwell G.M., Wurster C.F., and Isaacson P.A. (1967) DDT residues in an east coast estuary: a case of biological concentration of a persistent insecticide. *Science* 156: 821–824

Wright L.D. (1978) River deltas. In Davis R.A. (ed.) *Coastal Sedimentary Environments.* Springer-Verlag, Berlin, pp. 5–68

WWF-UK (1994a) Integrated coastal zone management: international commitments. *Marine Update no. 17.* WWF-UK, Godalming, November

—— (1994b) Local authorities and integrated coastal zone management plans. *Marine Update no. 18.* WWF-UK, Godalming, November

—— (1995) Integrated coastal zone management: UK and European initiatives. *Marine Update no. 19.* WWF-UK, Godalming, February

Xu Q.W. and Zheng J.L. (1991) Environmental management of the Bohai Sea in China. *Marine Pollution Bulletin* 23: 573–578

Yang G.S. and Zhu J.W. (1993) A study of impacts of global sea level rise on salt water intrusion into the Changjiang estuary. *Science in China Series B – Chemistry, Life Sciences and Earth Sciences* 36(11): 1391–1401

Yap H.T. (1996) Attempts at integrated coastal management in a developing country. *Marine Pollution Bulletin* 32(8–9): 588–591

Yapp G.A. (1986) Aspects of population, recreation and management of the Australian coastal zone. *Coastal Zone Management Journal* 14(1–2): 47–66

Yates B.F. (1994) Implementing coastal zone management policy: Kepulauan-Seribu marine park, Indonesia. *Coastal Management* 22(3): 235–249

Yohe G., Neumann J., Marshall P., and Ameden H. (1996) The economic cost of greenhouse induced sea level rise for developed property in the United States. *Climatic Change* 32(4): 387–410

Younger, M. (1990) Will the sea always win? Coastal management in north-east Norfolk. *Geography Review*, May: 2–6

Zann L.P. (1994) The status of coral reefs in south western Pacific Islands. *Marine Pollution Bulletin* 29(1–3): 52–61

Location index

Subject index